The Social Ecology of Tropical Forests

The Social Ecology of Tropical Forests

Migration, Populations and Frontiers

Edited by
Wil de Jong
Lye Tuck-Po
Abe Ken-ichi

Kyoto University Press

TRANS PACIFIC PRESS

First published in 2006 jointly by:

Kyoto University Press
Kyodai Kaikan
15-9 Yoshida Kawara-cho
Sakyo-ku, Kyoto 606-8305, Japan
Telephone: +81-75-761-6182
Fax: +81-75-761-6190
Email: sales@kyoto-up.gr.jp
Web: http://www.kyoto-up.gr.jp

Trans Pacific Press
PO Box 120, Rosanna, Melbourne
Victoria 3084, Australia
Telephone: +61 3 9459 3021
Fax: +61 3 9457 5923
Email: info@transpacificpress.com
Web: http://www.transpacificpress.com

Distributors

Asia and the Pacific
Kinokuniya Company Ltd.

Head office:
38-1 Sakuragaoka 5-chome, Setagaya-ku, Tokyo
156-8691, Japan
Telephone: +81 (0)3 3439 0161
Fax: +81 (0)3 3439 0839
Email: bkimp@kinokuniya.co.jp
Web: www.kinokuniya.co.jp

Asia-Pacific office:
Kinokuniya Book Stores of Singapore Pte., Ltd.
391B Orchard Road #13-06/07/08
Ngee Ann City Tower B
Singapore 238874
Telephone: +65 6276 5558
Fax: +65 6276 5570
Email: SSO@kinokuniya.co.jp

Japan
Kyoto University Press
Kyodai Kaikan
15-9 Yoshida Kawara-cho
Sakyo-ku, Kyoto 606-8305
Telephone: (075) 761-6182
Fax: (075) 761-6190
Email: sales@kyoto-up.gr.jp
Web: http://www.kyoto-up.gr.jp

UK and Europe
Asian Studies Book Services
Franseweg 55B, 3921 DE Elst, Utrecht,
The Netherlands
Telephone: +31 318 470 030
Fax: +31 318 470 073
Email: info@asianstudiesbooks.com
Web: http://www.asianstudiesbooks.com

USA and Canada
International Specialized Book
Services (ISBS)
920 NE 58th Avenue, Suite 300
Portland, Oregon 97213-3786
USA
Telephone: (800) 944-6190
Fax: (503) 280-8832
Email: orders@isbs.com
Web: http://www.isbs.com

Australia and New Zealand
UNIREPS
University of New South Wales
Sydney, NSW 2052
Australia
Telephone: +61(0)2-9664-0999
Fax: +61(0)2-9664-5420
Email: info.press@unsw.edu.au
Web: http://www.unireps.com.au

ISBN 1–920901–10–8 978–1–920901–10–3

Contents

List of Tables

List of Figures

List of Photographs

Abbreviations and Acronyms

CAR	Central African Republic
CARPE	Central African Regional Program for the Environment
DRC	Democratic Republic of the Congo
FAO	United Nations Food and Agricultural Organization
ha	hectares
IFRC	International Federation of Red Cross and Red Crescent Societies
IUCN	International Union for the Conservation of Nature
km	kilometers
NEP	New Economic Policy
NTFP	Non-timber forest products
PSF	Peat Swamp Forest
RP	rupiah
TREES	Tropical Ecosystem Environment Observations by Satellites
UNECA	United Nations Economic Commission for Africa
UNEP	United Nations Environment Programme
UNHCR	United Nations High Commissioner for Refugees
USAID	US Agency for International Development
WCMC	World Conservation Monitoring Centre
WRI-GFW	World Resources Institute-Global Forest Watch

Acknowledgements

The chapters in this volume were originally prepared for the International Forestry Research symposium, entitled "In, Out, In, Out: Populations, Migrations, and the Social Ecology of Tropical Forests," held in Kyoto from 5–7 December 2002, organized and co-sponsored by the Japan Center for Area Studies, National Museum of Ethnology and the Center for International Forestry Research, Bogor. We would like to especially thank the National Museum of Ethnology for its valuable support and the Japan Center for Area Studies for logistical assistance for the symposium and with the preparation of this volume. We especially thank Noriko Iizuka for her immense support during the symposium and with the preparations of this volume and Mayumi Tokuhira for help with the preparations of this book. We also thank Terry Rambo, Shigeo Kobayashi and Deanna Donovan, for sharing the sessions and leading the discussions at the end of the symposium. Thanks are due to Ms Miriam Riley for her editorial skills. Finally, we would like to thank the Japan Society for Promotion of Science for a Grant-in-Aid for Publication of Scientific Research Results, which enabled the publication of this volume.

Contributors to the Volume

ABE Ken-ichi is an associate professor of the Japan Center for Area Studies, National Museum of Ethnology. His areas of specialization are environmental anthropology and Southeast Asian region studies. He has studied tropical forests' political ecology in Southeast Asia, Chinese environmental history, and related subjects with the focus on the relationship between people and forests through repeated cycles of fieldwork. He organized several international symposia including, "UNESCO/JCAS Symposium and Experts Meeting: Exploring Linkages between Cultural Diversity and Biodiversity: Safeguarding the Transmission of Local & Indigenous Knowledge of Nature," 2005, and the session "Water and Cultural Diversity: Mediating for Sustainable Development," at the 4th World Water Forum, 2006. He has written and edited a number of books including *The Political Ecology of Tropical Forests in Southeast Asia: Historical Perspectives*, 2003, Kyoto University Press & Trans Pacific Press. He is also engaged in practical approaches including economic rehabilitation based on coffee cultivation in East Timor, where he had conducted fieldwork. Email: abek@idc.minpaku.ac.jp

Emmanuel ACHEAMPONG is a lecturer in the Department of Silviculture and Forest Management, Faculty of Renewable Natural Resources at the Kwame Nkrumah University of Science and Technology, Kumasi, Ghana. He received his MSc in Environmental Policy and Management and a PhD in Human Geography from the University of Hull, United Kingdom. After his PhD, he worked briefly as a Research Assistant in the Department of Geography, University of Hull. He also worked as a Sales Representative for Ghana Agro-food Company Limited before moving to Hull for his MSc and PhD. His research and teaching interests include: rural livelihoods and the relationship between communities and their natural resources; development issues at the people–environment interface; indigenous perceptions and valuations of environmental resources; and natural resource management. Email: achie70@hotmail.com

Leonard Y. ANDAYA received a BA in History from Yale University, and an MA and PhD in Southeast Asian history at Cornell University. He has held positions at the University of Malaya, the Australian National University, the University of Auckland, and the University of Hawaii, where he has been professor of Southeast Asian history since 1993. In 2000 he was a Senior

Scholar in the Fulbright-Hays program, which allowed him to conduct research in Southeast Asia, Europe, and Australia, and was a Fellow at the Netherlands Institute for Advanced Studies in Wassenaar, The Netherlands, in Spring 2000. Email: andaya@hawaii.edu

Phillip N. BRADLEY is Senior Lecturer in the Department of Geography, University of Hull where he has worked since 1995. Before then he was a senior research fellow at the Stockholm Environment Institute and prior to that he spent six years working in Kenya on development issues. He has worked extensively in Africa over the last thirty years, with a particular interest in rural livelihoods and the relationship between communities and their natural resources. A special concern is the changing circumstances of forest edge communities in the face of a declining resource base and the forces of modernization. Recent work in Zimbabwe and South Africa has focused on the prospect of shared resources between local communities and the state. This theme is continued in current work in the Atlantic rainforest of southern Brazil. He has also worked extensively on rural energy issues particularly the supply and consumption of wood fuels in Kenya, Zimbabwe and South Africa. This work has been closely linked with the changing role of forestry departments and their move towards social forestry. Email: p.n.bradley@hull.ac.uk

Carlos CORNEJO ARANA holds a degree in forestry engineering and an MSc in agricultural economics from the Universidad Nacional Agraria La Molina, Perú. He undertook postgraduate studies in enterprise management at the Universidad Nacional Mayor de San Marcos. Currently he is president of the Association Trópicos, a civil non-profit association that works with rural communities in the Peruvian Amazon, trying to establish new sustainable production alternatives based on the use of the biological diversity in the region. He is also the administrative director and associate of Biodiversidad Amazónica S. R. L., a private enterprise that conducts trials with the production of wildlife like capybaras and wild boars near Iquitos, applying productive strategies that include local rural communities. He has a special interest in the spatial occupation of territories in the Peruvian Amazon and its subsequent impact on natural resources. Email: ccornejo@iiap.org.pe

Wil DE JONG is Professor at the Japan Center for Area Studies, the National Museum of Ethnology in Osaka. He holds a PhD in Agricultural Sciences from Wageningen Agricultural University in Holland. His research deals with governance, economics and ecology of small holder natural resource management. Before he joined JCAS, he worked for nine years at the Center for International Forestry Research (CIFOR), with the headquarters in Bogor, where he directed research projects in Indonesia, Vietnam, Bolivia, Perú and Zimbabwe. Before CIFOR he worked for ten years as a Research Associate and an International Fellow at the Institute of Economic Botany at the New York Botanical Garden, conducting research in Perú and Indonesia. He has authored, co-authored or edited seventeen monographs, including two special issues of peer reviewed journals, twenty-nine peer reviewed journal articles, twenty book chapters, and some twenty non-peer reviewed articles and short notes. Email: wdejong@idc.minpaku.ac.jp

Edmond DOUNIAS is a scientist at the French Research Institute for Development (IRD, formerly ORSTOM). He has been working in Cameroon since 1984. He did his PhD research in coastal Cameroon, as part of an interdisciplinary program dealing with the anthropology of food and the bio-cultural adaptations of forest peoples to changing environments. In 1993, he received his French Doctorate with a specialization in ethno-ecology from the University of Montpellier 2. He did post-doctoral research granted by the L.S.B. Leakey Foundation on the ethno-botany of wild yams among the Baka Pygmies. He also coordinated a research program on the human ecology of forest-savannah boundaries in Central Cameroon. In September 2001 he joined the Center for International Forestry Research (CIFOR, Bogor Indonesia) and is now conducting fieldwork on the diet and health of the Punan forest dwellers of Borneo. Email: edounias@cgiar.org

Cristina EGHENTER held an Honorary Research Fellow position at the Centre for South-East Asian Studies, University of Hull, UK, and worked with WWF Indonesia, Kayan Mentarang Project, as Director of the Culture and Conservation program (1995–1997), and Coordinator of Community Empowerment (1999–present). She received her M Phil and PhD in Anthropology from Rutgers University, USA. She was a post-doctoral fellow (1997–1999) at the Centre for South-East Asian Studies, University of Hull, and Visiting Research Fellow at the International Institute for Asian Studies (IIAS), Amsterdam, and the Institute of Southeast Asian Studies (ISEAS), Singapore. Her main fields of interest include: environmental anthropology; conservation and development issues, natural resource management, knowledge and cognition; explanation in the social sciences; and Borneo ethnography. Email: awing@samarinda.org

Christian LECLERC is a scientist at the French Agricultural Research Centre for International Development (CIRAD). He has been working in

Cameroon and Gabon since 1994 among the Medjan and Baka Pygmies. He was a member of the Future of Rainforest Peoples Project (FRP) financed by the DG VIII of the European Union (1995–2000). In 2001, he received his French Doctorate with a specialization in ethnology and comparative sociology from the University of Paris. He is now conducting research on the in situ management of genetic resources in Africa in order to better assess the interactions between the biological and the socio-cultural dynamics that contribute to the conservation and the valorization of crop genetic diversity. Email: christian.leclerc@cefe.cnrs.fr

Luis LIMACHI HUALLPA is an economist who graduated from the National University San Antonio Abad in Cusco, Perú. He did his Masters at the Pontifical Catholic University in Lima, as well as postgraduate courses within Perú and abroad. Luis has lectured at the National University San Antonio Abad, and at the Peruvian Amazon University. Currently, he works as a researcher at the Peruvian Amazon Research Institute (IIAP). His main responsibilities are characterization and spatial representation of socio-economic aspects of the Peruvian Amazon. Email: llimachi@iiap.org.pe

Víctor Eduardo LLANQUE ZONTA currently studies anthropology and economics at Macalester College, Minnesota. During his previous studies at the United World Colleges, he focused on development studies. While working at the Instituto Para el Hombre la Agricultura y la Ecologia of Riberalta Bolivia, his interest was on sustainable development efforts in the Bolivian Amazon. As the "Immigrant and Refugee Issue Area Coordinator" at Macalester College's Community Service Office, his current work and research focuses on culture, ethnicity, immigration, and economic justice in the United States. Email: vllanquezonta@macalester.edu

Stephen G. PERZ received his PhD in sociology with a specialization in demography from the University of Texas at Austin in 1997. Perz conducts research on demographic processes and environmental change, land-use practices and land-cover dynamics, and the socio-spatial processes of road-building in the Amazon. He has received funding from NASA and NSF for work on roads and forest fragmentation in the Brazilian Amazon. He is involved in fieldwork along the Transamazon highway that crosses northern Brazil, as well as along the Transoceanic highway that connects southern Brazil to Perú and Bolivia. This work emphasizes interdisciplinary and international collaboration and has resulted in roughly thirty publications in numerous journals and books. Email: sperz@soc.ufl.edu

Lesley POTTER is Associate Professor in the Department of Geographical and Environmental Studies, University of Adelaide. She received her MA from the University of Melbourne and her PhD from McGill University, Montreal, and has taught at the University of Guyana (South America) and the University of Canterbury, Christchurch, New Zealand. She has worked in Southeast Asia for over twenty years and has carried out many field-based studies, in Indonesia, Vietnam, Laos, and Malaysia, together with archival research covering the whole of Southeast Asia. She has published more than forty book chapters and articles in international journals, and one jointly authored book. Email: lesley.potter@adelaide.edu.au

LYE Tuck-Po is an environmental anthropologist affiliated with the Naga Research Group. She received her graduate degrees at the University of Hawai`i at Manoa, where she was also a Degree Fellow at the East-West Center. Previous positions include the Quillian Visiting International Professorship at Randolph-Macon Woman's College in Virginia and fellowships at the Japan Center for Area Studies, the Center for Southeast Asian Studies, Kyoto University, and the Research School of Asian and Pacific Studies, Australian National University. She has researched the environmental knowledge and landscape perceptions of Batek hunter-gatherers in Malaysia, and currently focuses on land and water management among peasants in Cambodia. In addition to two edited volumes and several articles, she has published *Changing Pathways: Forest degradation and the Batek of Pahang, Malaysia* (LexingtonBooks, Lanham, MD). Email: tuckpo@streamyx.com

TRAN Van Con is director of the Tropical Forest Research Centre of the Forest Science Institute of Vietnam. He graduated as a forest engineer at the Technical University of Dresden (Germany) in 1979 and received his PhD in silviculture from the Forest Science Institute of Vietnam (1991). His fields of interests are: forest inventory; agroforestry; social forestry; and community development. He has published studies on community forest management and swidden agriculture. Email: tfrc@dng.vnn.vn

1
Migration and the Social Ecology of Tropical Forests

Wil de Jong, Lye Tuck-Po and Abe Ken-ichi

Introduction

The specter of landless migrants moving into tropical forests, clear-felling as they go, causing irreversible damage to once-pristine areas, is one of the common images of our time. A plethora of studies have proliferated as researchers try to uncover the reality behind these images (for example Bilsborrow 2002; Perz 2002). Much of this literature is primarily concerned with tracing the causes of deforestation. A related strand of studies pays more attention to the process of migration itself. We define migration as the spatial relocation of people from one place to another, whether regionally (within national borders) or internationally (across national boundaries). Is migration a direct cause of deforestation in the tropics or is it one intervening factor among many? If the latter, then what are the indirect causes urging migrants to move into forested lands? And what happens once they arrive? Is deforestation a necessary consequence?

These are some of the questions considered in the present volume. We provide an overview of the linkages between migration, migrants, long-term forest dwellers, and forest ecology in the three major tropical regions (America, Africa, and Asia). Our purpose, however, extends beyond tracing these connections. We examine the social ecology of migrants in a forest environment in order to better understand forest-cover change. Our discussion is equally motivated by an interest in the processes involved in tropical forest migration itself. Although, as we discuss in this chapter, migration into forested areas is as old as human history, and accordingly has drawn the attention of academics from many disciplines, the environmental impact of migrants in a diminishing tropical forest frontier is still not fully understood.

For these purposes, we present in this volume a diverse set of studies on migration, migrants and tropical forests. The chapters have been grouped regionally, opening with Latin America, followed by Africa and finally

Southeast Asia. Each of the three sets of studies begins with an introductory chapter followed by specific case studies that focus on local areas and regions. In this chapter, we supply a succinct overview of the topic of migration in relation to tropical forests, discussing some of the main themes addressed in the scientific literature. An analysis of the concepts of "migration" and "social ecology" follows. We then discuss the chapters of this volume, summarizing their contributions to the discourse on migration and the social ecology of tropical forests.

A brief overview of migration and tropical forests

For centuries, tropical forests have served as a frontier, drawing those in search of fabled abundant natural resources, rulers seeking new territories or manpower for armies, refugees escaping wars and skirmishes, and even slave traders. Both regional and international trade brought Europeans to the Americas. They were soon drawn to the Amazon forest as they sought new forest products or land for agriculture. Prior to this time, the spice trade had attracted first Chinese, Arabic and Indian traders and then European powers to Southeast Asia to exploit natural stands of clove, nutmeg and pepper. The Europeans vied for the monopolization of trade and were responsible for the earliest attempts to establish crops in large estate plantations.

Preceding and during these international movements of traders and colonizers, regional and local migrations were occurring. Although the majority of these waves of migration had an economic basis, many were part of local struggles for power and territorial control (see for example, Adas 1992; Sellato 2001; Vickery 1998). The international trade in forest products, the occupation of foreign lands by colonial powers, and the impact of these factors on local cultures have all been the subject of research.

Those in power have played a distinct role regarding tropical forest migration. In general, conflict between planned and spontaneous migrations has been the primary results of these influences. Rulers in pre-colonial Asia and Africa based their powers on the control of people rather than land (see Adas 1992). Hence, while they at times initiated expeditions to conquer peripheral peoples, they also had an interest in preventing subordinates from moving out of their spheres of control. European rulers encouraged their peoples to seek, exploit, and claim forest resources in remote countries in the name of the Crown. Upon colonization they discouraged the spontaneous movements of indigenous groups that threatened their territorial control. At the same time, colonial powers initiated the first planned relocations of people (for example, the importation of foreign laborers to work in plantations and mines, the relocation of Javanese to Surinam by the Dutch, and the relocation of indigenous farmers by the British in Africa to free the land for lucrative commercial agriculture). With the growth in the international timber trade from the nineteenth century onwards, colonial powers became concerned about the environmental impact of forest-related migration (Potter 2003). Thus, although Dutch and British administrators remained unconcerned about the environmental impacts of plantation-based forest clearing (Kato 1991), they strove to prevent the movements of forest dwellers (especially swidden farmers) on the basis that their incessant clearing would destroy the forests. The Dutch and British engaged in wide-scale forest clearing and imported migrant labor to do so, but condemned spontaneous indigenous migration for its affects (usually construed negatively) on tropical forests (Potter 2003).

The tropical forest–migration link continues today, exhibiting some of the same themes outlined above. Wars in places such as Central Africa continue to push refugees deeper into forests while in other regions such as Indo-China, the cessation of war has meant that new settlements can now be established in previously restricted areas (de Jong *et al.* forthcoming). Agricultural frontiers in many regions of the tropical world are closing, thus exacerbating the need for new arable lands. Resources other than timber, such as valuable minerals (gold, diamonds, coal, oil, etc.) and even botanical genetic resources continue to draw multinational operations and their employees. Since plantation crops are a major source of revenue for many countries, deforestation is occurring on a large scale. Simultaneously, the decline of forest ecosystems is placing unprecedented stresses upon the support systems of previously sheltered forest dwellers.

These pressures, coupled with governmental exhortations to "modernize", foster amongst forest dwellers the desire and necessity to seek better social and economic opportunities elsewhere. Migration *out* of forests in the direction of urban areas is therefore ongoing. Alongside the reality of declining forest communities lies another; individuals, and even entire communities, are relocating within forest territories. These "internal migrations" are perhaps motivated by a variety of reasons, including cultural ethos (for example when movement underpins a way of life, as for hunter-gatherers and swidden farmers), local resource shortages, and government-sponsored resettlement projects. As this brief summary shows, migration has far-reaching social and ecological consequences. Migration affects people residing both in the destination and origin of migration.

Since the 1980s forest-related migration, in conjunction with concern over tropical forest decline, has increasingly received attention from academic, governmental, and conservationist discourses. Contributors to these discussions have linked contemporary migration to the wellbeing of forest-dependent peoples (Sader *et al.* 1997). Simplistic explanations hold that migration causes accelerated population growth that leads to pressure upon resources and deforestation. In academic discourse it is argued that migration is an intermediate result that is caused by underlying factors such as infrastructure development, economic and fiscal policies and trends, or regional wars. During the 1990s studies focused on the push- and pull-factors causing migrations rather than analyzing the impacts of increasing numbers of people on forest ecologies. For example, Kaimowitz, Thiele and Pacheco (1999) examined the links between structural adjustment programs, migration, and deforestation in Bolivia, finding that macro-economic reforms remove incentives for some sectors to move to the forest for agricultural purposes, but increase the need for others to turn to the forest frontier in search of replacement for lost economic opportunities. The displacement of rural farmers as a result of economic liberalization and reform also leads to widespread forest frontier migration (see Browder and Geoffrey 1997). Pokam and Sunderlin (1999) and Sunderlin and Pokam (2002) argue that the economic crisis in Cameroon is the primary cause of migration to the forests.

Although deforestation itself is a reality, many studies assume a simple correlation between the statistical analysis of forest loss and population increase without considering the processes involved in this relationship. An exception is the study of forest conversion in Indonesia by Sunderlin and Resosudarno (1999). They argue that population pressure is an intermediate factor therefore its contribution to deforestation depends on additional parameters. Migration is the result of underlying factors. The environmental impact of migrants in their new locations is also conditioned by a number of external factors, such as the price of agricultural commodities. Several chapters in this volume (Perz, Chapter 2; Llanque, Chapter 4; and Limachi *et al.*, Chapter 3) echo this finding.

State-sponsored migration continues to be a phenomenon in all three tropical regions. The most widely-known are the transmigration schemes in Indonesia (Swasono and Singarimbun 1986; Fasbender and Erbe 1990; Levang 1997; Potter, Chapter 8), and the various programs initiated by the Brazilian government to open up the Amazon basin (see Browder and Geoffrey 1997 for an overview). Often these state-sponsored schemes trigger spontaneous migration as new developments and economic opportunities appear (see Charras and Pain 1993; Browder and Geoffrey 1997; Poffenberger 1998). These in turn lead to spiraling levels of social and environmental change. As such, several studies have advanced beyond looking at only the projects, instead focusing on the impacts of these initiatives on the environment, local peoples, and transmigrants themselves (see Abdoellah 1993; Tran, Chapter 10).

A subsequent group of studies highlights the negative social and economic impacts of forest frontier migration (see Subedi *et al.* 1993). The frontier, as defined by Frederick Turner (1893), is "the outer edge of the wave—the meeting point between savagery and civilization". The forest frontier is where migrants and locals are brought together, sometimes resulting in conflict, and often leading to unpredictable social, cultural, economic, and environmental outcomes. In-migrants may or may not be ethnically similar to the locals. Certainly, they bring with them new customs, technologies, and needs. A common result of migration is profound changes to existing land ownership patterns (see Ayuwat 1993; Tran, Chapter 10; Acheampong, Chapter 5).

However, the frontier is also a place of reunions. Sometimes migrants are *reverse migrants*, that is, people who return to their place of origin after having been away for some time (Blahna 1990; Eghenter, Chapter 11). Returning home, such migrants may initiate new endeavors, but these often fail, forcing the returnee to change activities or move on elsewhere. Changes in prices of products grown in or harvested from the forest—the resources that initially motivated migration—force migrants to find new sources of valuable products (for example eaglewood: see Eghenter, Chapter 11), turn to other cultivars (for example, switching from rubber to brazil nut: see Llanque, Chapter 4), or may cause serious social stresses and conflicts (for example, in the case of the decline of coffee prices in Vietnam: see Tran, Chapter 10; Potter, Chapter 8).

Researchers have studied indigenous groups that inhabit tropical forests and their patterns of migration (for example Eghenter 1999; Padoch 1982). Only recently have detailed analyses appeared of how these migration patterns have influenced spatial occupation and resource control or contestation (for example Sellato 2001; Kaskija 2002; Dounias and Leclerc, Chapter 7). Contemporary spatial distribution of indigenous groups of East Kalimantan, for instance, has been explained by the control of areas rich in forest resources such as birds' nest caves and eaglewood that attract a high level of international demand.

The examples discussed so far only include migration into and within tropical forests. An extensive literature analyses the alternative scenario: migration *out* of the forest and into urban areas (Brunson and Kennedy 1995; Lohnert and Geist 1999). This genre of studies first appeared as a result of the focus on rural development in newly independent countries that occurred after World War II. The effect of this type of migration on tropical forests is distinct and may potentially relieve pressure on forests. However, often the effects are equally ambiguous from the perspective of forest conservation (Llanque, Chapter 4). For example, the flow of peoples out of forests may strengthen forest-dwellers' links to the market economy, providing new sources of demand for forest products.

Migration and migrants

As the foregoing overview suggests, the definition of migration that we provided at the outset—as the spatial relocation of people intra- and inter-regionally—does not capture the complexity of the process. Migration implies "changing places" (Ingold 1986) but so does many other kinds of movement. One can think of various examples, ranging in scale and magnitude from shifting one's home to a different part of a small village to full-scale emigration overseas. Further, migration does not imply a particular direction of movement: the examples that we cover in this volume also include moving back (reverse migration), moving around (circular and seasonal migrations), and moving on (temporary migration). Cultural factors influence the style of migration undertaken. For refugees, or more generally those who are forced to move, migration is arguably only the means to an end; these people would prefer not to relocate and, once moved, hope to settle down as expeditiously as possible. For historically mobile populations, such as hunter-gatherers, pastoralists, maritime peoples, and swidden agriculturists, movement may be an end in itself (Lye 2004).

Migration styles have environmental consequences. Whether migrants move within their own territory or to alien terrain, their knowledge of local ecology is affected. Their ability to exploit the environment successfully depends both on having prior ecological knowledge and the skills to access local knowledge (Eghenter, Chapter 11). Location of movement also affects how migrants view and use their environment: whether it is a long-term settlement therefore requiring proper regulation and resource management, or a temporary relocation fostering rapid exploitation before a move can be made to more salutary surroundings.

In view of these complexities, we need to look more closely at the characteristics of migration, migrants, and the environments they inhabit. We propose three axes of differentiation: the distance of movement, the reasons underlying movement, and the duration of movement. We then propose a rough taxonomy of tropical forest migrants.

Distance of movement

When we look at the types of movement that people engage in, some exclusions from the definition of migration are apparent. Mobile populations who move relatively short distances within their own territorial base are a clear exclusion. Movements in these examples are usually patterned according to social and environmental logic, such as following seasons or lifecycle events, and follow well-defined routes (Lye 2004; Sellato 1994; Sutlive 1978). Moving to a new place often does not imply abandoning past settlements and resource bases (Silberbauer 1994). Nor would we include in this category the large numbers of people from Europe, the United States, Japan, and other affluent countries who engage on annual or biannual holiday travels in far-off countries. While for both tourists and mobile peoples movement may be part of their way of life, it does not imply movement out of a home community.

Most people would consider migrants to be people who move a considerable distance, that is, at least from one region to another within a country. Yet, as the above exceptions show, distance is not the defining factor. There is also the question of whether the migration represents a significant departure from other kinds of movement that people engage in. Further, the reasons why people move out, and how long they stay once they arrive, appear to be important criteria. In conclusion, migrants and migration are defined by the features of the movement itself, and, as we demonstrate below, by the reasons motivating the migration.

Reasons for movement

The reasons why people relocate are many. Causality is possibly the key feature to differentiate the various types of migration. Migration may be voluntary or involuntary. Involuntary migration includes forced relocation or flight from wars. It may also stem from government-sponsored resettlement (for example to make room for plantations, protected areas, and infrastructure development). Even in cases where there is no coercion involved, people may feel forced to move as resource bases decline and frontiers open up elsewhere. Voluntary migration often has an element of opportunism, especially when movements are triggered by resource booms. Often migrants, whether they move to a different country or to forest frontiers, are attracted to new opportunities for improving their material wellbeing. Historically, forestlands have been viewed as empty lands that could be cleared for production purposes (Lim 1976). This perception is still evident in the frontiers.

How we classify an individual as a migrant is less obvious. For governmental purposes, there is a clear bureaucratic demarcation between economic migrants and political refugees (or exiles). Further, a state-centric perspective often distinguishes between legal and illegal immigrants, with the latter often classified as "squatters". Such distinctions are often arbitrary and may have less to do with actual criminality than with the compatibility between migrant behaviors and governmental projects. As we discussed above, state administrations, whether pre-colonial, colonial, or postcolonial, often restrict spontaneous migration. However, when it is in their interest to do so, they have in fact created new migrant frontiers. Many would argue that refugees who flee from their countries are migrants or become migrants as their needs for sustenance and support are similar to that of other types of migrants (see Bradley, Chapter 5). Swatuk (forthcoming) and Price *et al.* (2004) instead place refugees in a separate category as their reasons to relocate are different: migrants move for economic reasons, while refugees move because they fear persecution and violence. Although the distinction between the two groups is useful, both often end up at forest frontiers where they interact with the environment, the local people, and each other. As such they fall within the focus of this book.

Duration of movement

A third defining characteristics of migration is the duration of the movement. In general, migration implies that people relocate for some minimal duration. For Perz (Chapter 2), migrants are defined as people who arrived in a place during the five year period prior to a given census, thus assuring comparability among migration patterns in the countries that he compares. However, this methodology is not comprehensive: what, for example, of the case of children who are born after their parents' migration and have not developed social skills in their parents' place of origin? Further, migration often leads to the permanent establishment of populations. In such cases it may be possible to distinguish migrant communities from indigenous peoples (if these exist in the area), but the line between migrants and older waves of settlers is by no means clear-cut.

As outlined earlier, migration is not limited to unidirectional movement, but can be bi- or multidirectional. People may relocate for a limited amount of time, then return home or move on elsewhere. Second generation migrants (such as in plantation-based communities) may move away. In circular migration, workers move according to seasonal employment; they neither abandon their home communities nor fully settle down in their place of emigration. In this case it is the migration pattern itself that endures rather than actual residence in a particular community.

Despite its problems, duration of stay is a useful criterion. It provides information about the long-term residence choices and patterns of the migrants themselves. It could be argued that mobile loggers and entrepreneurs who enter the forest to engage in forest product exploitation are not migrants because they do not intend to occupy their resource zones permanently. In this volume, however, we include chapters that discuss these short-term relocations as they are often linked to other migration patterns that have already occurred or are occurring simultaneously (see Llanque, Chapter 4; Eghenter, Chapter 11).

A brief taxonomy of migrants

Having examined some of the standard patterns of migration and the exceptions to these examples, we recognize that no taxonomy will ever capture the complexities involved in the migrant–forest relationship. However, it is necessary to briefly highlight several key aspects of this dialectic.

One criterion of this taxonomy is the environments of the migrant's origin and their destination. Using this measure, it is possible to identify five types of migrants: forest–forest migrants, migrants leaving the forest, rural–urban and urban–rural migrants, and migrants who move from one country to another. Each of these situations can be further specified (for example, migrants moving from forest interiors to forest margins). This typology then allows us to frame research questions on the potential impact of migration on particular types of environments, providing insights into the issue of sustainability.

Further determinants are the reasons motivating migration. Several of these causes have been summarized above: economic (likely to be a dominant reason), cultural and kinship imperatives, government resettlement, and flight from armed conflicts, oppression or persecution. Elucidating motivation may be problematic, since motives are often mixed. This process is useful, however, as it provides an understanding of what migrants' needs and aspirations may be, and therefore the types of resource management they may employ.

Several subgroups of migrants can be distinguished, for instance seasonal migrants (a subgroup of economic migrants), frontier pioneers (economic, sometimes cultural migrants), and returnees (mostly refugees, but sometimes economic migrants). These subdivisions can be extended for particular purposes. In the case of forest-related migrants, the long-term impact on the forest can be further utilized to typologize migrants. One can distinguish, for instance, between migrants who settle, build houses and create their own estates by converting forest into agricultural lands, and migrants who follow economic imperatives. The latter will remain only as long as the opportunities exist that support their trading interests.

Social ecology

The term social ecology was first used in the analysis of the patterns of behavior of animal populations. In the contemporary literature, these studies would fall within the field of population ecology. In more recent literature, the term "social ecology" has taken on a diverse range of uses. For example, the term has been used both in pediatric psychology (Brown 2002) and the social psychology of medicine (Tickle-Degnen and Doyle Lyons 2004), to describe the behavioral process between children in the case of the former and patients in the latter. These examples employ the term social ecology in the investigation of interactions between organisms and their environments. Since the mid-1980s, however, the term has increasingly been used in scholarly discussions to describe links between social phenomena and the ecology of natural resource use.[1]

We propose the term "social ecology" as a complement to two other related analytical frameworks: "human ecology" and "political ecology". Linking human and social ecology is the study of adaptation in terms of responses to environmental opportunities and constraints (Ellen 1982; Moran 2000; Rambo 1983). Human ecology has its origins in classical anthropology and focuses on the domain of interaction between social actors and their surrounds. This approach has greatly contributed to the understanding of the relations between humans and tropical forests (see, for instance, Rappaport 1967). The scope, however, of the field of human ecology can be limiting when employed to explain the dynamics between peoples and their natural environment within a wider economic, political and social context. This limitation led to the rise of political ecology, a field that draws upon the theories and approaches of political economy to explain environmental relations (Bryant 1992; Bryant and Bailey 1997; Greenberg and Park 1994; Peet and Watts 1996). Political ecology looks beyond people and their immediate environments to analyze the ramifying effects of extraneous and wide-ranging factors such as market-based instruments for change, global trade, fiscal and economic policies, and the role of key players in the upper echelons of the capitalist economy (Lye *et al.* 2003). Many political ecology studies focus on the role of power relations in resource use and management, in particular, coercive power (de Jong *et al.* 2003).

We propose use of the term social ecology, as neither human ecology nor political ecology provide adequate frameworks to facilitate analysis of processes such as migration into and out of tropical forests. The human ecology framework could be utilized to explain how people adapt to environmental constraints and opportunities in forest frontiers, but not how those constraints and opportunities are shaped by outside forces (see Llanque, Chapter 4; Dounias and Leclerc, Chapter 7; Tran, Chapter 10; Eghenter, Chapter 11). Political ecology fills this gap by affording analysis of the dynamics of external forces (players, politics, and policies, and the relationships amongst them); however, this framework is perhaps limiting by its focus on these forces. Social ecology has a wider compass. It establishes and emphasizes the ecological impacts of migration, also taking into account the social effects of the forest environment on migrants and the macro-social effects of migrant–frontier interaction. Social ecology is both a description (the social ecology of migration) and an approach (the study of migration as a social phenomenon with multifaceted implications).

Converging themes

The chapters in this volume reflect a geographical and historical range of analyses on the theme of tropical forest migration. The discussion encompasses the main tropical forest regions: the Amazon (Chapters 2–4), Central and West Africa (Chapters 5–7), and Southeast Asia (Chapters 8–12). These take into account temporality—from the early modern (Andaya, Chapter 9; Potter, Chapter 8) to the contemporary period. Several chapters cover both historical and contemporary themes (Llanque, Chapter 4; Acheampong, Chapter 6; Limachi *et al.*, Chapter 3). The chapters by Perz, Bradley, and Potter summarize the key themes concerning the Amazon region, Central West Africa and Southeast Asia respectively, while other chapters focus upon a specific country or region. Only the paper by Dounias and Leclerc (Chapter 7) attempts a cross-continental comparison.

Forest conversion is a recurring theme in this volume. Perz (Chapter 3), and Limachi *et al.* (Chapter 4) make contemporary forest conversion processes in Bolivia, Brazil, Ecuador and Perú the primary foci of their discussions, while Potter (Chapter 8) reviews migration and forest conversion since the late nineteenth century in Southeast Asia. Several other contributors, like Acheampong (Chapter 6, Ghana), Bradley (Chapter 5, Western and Central Africa), Andaya (Chapter 9, pre-colonial Southeast Asia) Tran (Chapter 10, Central Highlands of Vietnam), and Llanque (Chapter 4) discuss forest conversion in relation to specific cases of migration. However, these chapters also analyze other issues relevant to tropical forest-related migration.

The three chapters on Latin America, that deal with Bolivia, Brazil, Ecuador and Perú (Perz, Chapter 2; Llanque, Chapter 4; Limachi *et al.*, Chapter 3) each pursue a distinct focus, despite the fact that migration and forest conversion feature prominently in two of the three chapters. Perz (Chapter, 2) shows that recurring and dominant land-use patterns and the national political economy help to explain observed statistical links between migration and forest cover. Limachi *et al.* (Chapter 3) present a comprehensive overview of processes of migration into, out of, and within Perú's tropical forest region, linking these trends with national policies and major social events, as well as national and international trends in agricultural commodity prices. The general patterns observed are supported by a rich amount of data on people moving from the various regions, and with data on forest-cover change. Llanque (Chapter 4) adopts a very different line of inquiry as he analyses migration trends and their consequences in a region that has had an economy entirely based on forest product extraction for over a century. This region has experienced in-migration, out-migration, rural–urban, urban–rural and rural–rural migration. Different external factors explain each of these types of migration, and each of these factors has had a significant impact on many aspects of life.

Several chapters in this volume detail historical cases of tropical forest migration. Andaya (Chapter 9) discusses pre-modern migration in Sumatra, Peninsular Malaysia, and Borneo, showing that forest-related migration and some of the political and economic forces behind it have been continuous since pre-colonial times. Potter (Chapter 8) continues the history of migration in Southeast Asia from the point at which Andaya's account concludes. Both these chapters, and particularly Potter's, are thematically tied to the migration–deforestation chapters, discussing forest conversion related to migration since the late nineteenth century. Bradley (Chapter 5) questions the presumed migration–deforestation link in Africa during the same period, arguing that much of the data that has been used to draw this conclusion is unreliable.

Both Eghenter's (Chapter 11) and Dounias and Leclerc's (Chapter 7) chapters briefly discuss historical migration in Borneo, although the former largely as an introduction to more recent migration trends and the latter as the basis for a theoretical discussion. These authors, together with Bradley (Chapter 5) and Acheampong (Chapter 6), contribute to a detailed review of various historical trends. The Latin American chapters rely upon historical analysis to a lesser degree. Llanque (Chapter 4) and Limachi *et al.* (Chapter 5) make reference to some of the historical trends that contribute to a better understanding of the contemporary cases they present, but history is not a major focus of these chapters.

A final converging focus of the chapters in this volume are the social, cultural and economic outcomes of processes of migration as outsiders move into forested areas. This is discussed by Acheampong for Ghana (Chapter 6),

Tran for the Central Highlands in Vietnam (Chapter 10), and Eghenter for East Kalimantan (Chapter 11). Specific themes that these chapter draw out include the dynamics of property rights, strategies employed by immigrants to deal with the social barriers that they encounter, and emerging and real conflicts.

The social ecology of tropical forest migration

The field of social ecology raises a number of questions that can be asked in order to provide a comprehensive view of the link between migration and the social ecology of tropical forests. The most obvious questions are: Are there historical trends? What are the experiences of migrants once they arrive at new forest frontiers? How do migrants affect the local population? Conversely, how does the new forest environment and its inhabitants affect the migrants? This section provides a brief summary of the answers to these questions provided in this book.

When and why?

Several of the chapters in this volume provide examples of migration from the early trade exchanges between European countries and tropical regions. Bradley (Chapter 4) observes that the humid forest zone of Africa has witnessed constant movement of peoples for the past 500 years (see also Dounias and Leclerc, Chapter 7). Andaya (Chapter 9), referring to Southeast Asia, observed that the early pepper production and tin mining initiatives led to both forced and spontaneous migration to the areas of production. The key point here is that significant population movements in Southeast Asia actually precede colonization.

Several of the authors point out that the rise of mercantilist activities has played a major role in tropical forest-related migration. These mercantilist activities, however, had different forms and subsequently different effects on migration patterns. Acheampong (Chapter 6), for instance, explains that the Akan, one of the pre-colonial groups in what is today Ghana, began trading gold in the fifteenth and sixteenth centuries. This trade enabled these groups to import slaves utilized to expand agricultural activities. The new centers of agricultural production that resulted from this sequence of events attracted new farmers that became interested in the expanding activities in these zones. This is one of the origins of accelerated forest frontier expansion in this part of Africa.

The phenomenon of slave raiding, and its profound impact on peoples and communities, is reported by Dounias and Leclerc (Chapter 7) and is covered in less detail in other chapters. They discuss the example of the Aka pygmies in the moist tropical forest zone of western Africa. The contemporary distribution of these people is the result of, on the one hand, their flight from slave raiding indigenous groups of that part of Africa, and, on the other hand, their alignment with groups fleeing from slave raiders themselves. Some of the contemporary locations of the indigenous groups of Eastern Borneo are at least in part attributable to hiding from slave raiders (Dounias and Leclerc, Chapter 7; Kaskija 2002; Sellato 2001; de Jong *et al.* 2003). Massive movements of indigenous groups that were pushed into the interior of tropical forests by other indigenous groups, as the latter sought slaves or new territory, has been a common feature in the South American tropical rainforest belt (Denevan 1976; Meggers 1971; Lathrap 1970).

The picture changed significantly once plantation development started in both Africa and Asia. The early example of plantation development and migration is the pepper production by local rulers that occurred in various parts of mainland and insular Southeast Asia. Pepper was heavily traded from the early sixteenth century. Local rulers widely encouraged groups to relocate to their territory. They also bought slaves and sent ships to bring in forced workers (Andaya, Chapter 9).

Once colonial powers had established trading interests and monopolies in Asia and Africa, market-based plantation production began. This continued throughout much of the nineteenth and well into the twentieth centuries. Before the colonial rulers in Africa turned to crop production, they forced indigenous Africans to extract minerals, timber, rubber and other forest products, and consequently to migrate to the areas where these products could be found. Since the late nineteenth century, however, African colonial powers also turned to crop production, thus requiring a substantial labor-force (Bradley, Chapter 5). Following the cultivation of pepper and other spices, coffee and eventually rubber became the most important plantation crops in Asia. As these plantation industries developed, plantation owners turned to contracting laborers. Since 1870, the plantation industries in Sumatra have imported large numbers of contract laborers from Java and other parts of Indonesia. Importantly, once contract laborers had signed up and arrived, their lives were little different from that of forced laborers. Rubber plantations in Malaysia (then Malaya) brought in contract workers from distant India in the first decades of the twentieth century, while rubber plantation owners in Vietnam imported workers from Tonkin (Potter, Chapter 8). In each of these examples, the workforce was relocated to forest frontiers where laborers engaged in clearing the forest to establish and then tend the plantations.

Some exceptions to this export-driven pattern are reported in the chapters of this book. One example of this is the migration of the Iban of West Borneo, a migration based on a cultural desire to pioneer new areas and expand control over unoccupied lands but also related to the self-image of males—experience in migration and seeking wealth elsewhere was one means to claim leadership qualifications. Indeed, this culturally induced migration led to an expansion from the mid-sixteenth century onwards from the Kapuas region, in what is today West Kalimantan, Indonesia, into the Second and Third Divisions of Sarawak (Andaya, Chapter 9). Another example is the occupation and opening up of the Irrawaddy delta in lower Burma. Participants in this migration arrived largely on their own initiatives into this part of the country from northern Burma after periods of droughts and food shortages there. People were driven by the need to produce their own food, but also by the opportunity to produce rice for sale in international markets. Government support and opportunities to achieve assured tenure assisted the initially itinerant migrants to remain. Arguing along similar lines, Eghenter (Chapter 11) summarizes some of the reasons why the Kenyah moved to the Apo Kayan before control was imposed on their migration by successive administrations of the Dutch and Indonesian governments. The Kenyah moved into, within, or away from the Apo Kayan because of inter-group warfare, leadership disputes, religious beliefs, and demographic and environmental changes.

Another type of migration relates to the exploitation of naturally occurring forest products. Migration related to product *extraction* implies a different kind of mobility that is linked to the ecology of the product. It also has very different implications for the effect of migration upon forest condition. This issue is mentioned in several of the chapters in this book, and two contemporary examples of this are discussed by Llanque (Chapter 4, brazil nut exploitation in northern Bolivia) and Eghenter (Chapter 11, eaglewood in East Kalimantan) respectively. One of the most prominent examples of product extraction is rubber exploitation in the Amazon basin. Llanque (Chapter 4) briefly refers to this case as a prelude to his discussion of modern migration into northern Bolivian forests. Bradley (Chapter 5) refers to the cases of minerals, timber and rubber, while Potter (Chapter 8) mentions the case of gutta percha (a tree resin used among others for cable coating) extraction between 1870 and 1910—exactly the same period that the rubber boom brought tens of thousands of fortune seekers to the Amazon. The availability and demand for rubber attracted large numbers of local as well as foreign migrants to the regions where natural stands occurred in Bolivia, Brazil, Perú, and Colombia. Gutta percha was very different, as it was less available and less in demand. Like many similar forest products from Southeast Asia, it was procured largely by indigenous people such as the Iban in Borneo and the Orang Asli in Peninsular Malaysia (Dove 1994; Potter, Chapter 8). Both rubber and gutta percha went through a similar production history and were eventually established as plantation crops.

A period without forest-oriented migration

While the period between the end of the nineteenth and early twentieth centuries experienced a boom in migration to forested areas, much of the first half of the twentieth century appears to have been relatively quiet in this respect. There are several reasons for this. Towards the end of the 1910s, natural rubber extraction in the Amazon basin declined to a minimal level in comparison to the boom period of the previous twenty years. Large numbers of rubber boom migrants left the forest. Subsequent minor booms in the Amazon never reached the same magnitude (San Roman 1975).

In Asia, the plantation era, and the contracting of large numbers of workers from afar, declined drastically in this period. In addition, the Dutch colonial government did all possible to stop the sort of migration as described for the Kenyah by Eghenter (Chapter 11). As a result of pacification and Christianization, warfare and religion became less significant triggers of migration. At this time Africa also experienced a decline in plantation production. An additional factor that has reduced the opportunity to migrate within Africa is the wave of nation-state independence that took place from the 1950s onwards. Countries became more aware of territorial integrity and began closing borders. This development affected migration across borders, though it was never stopped completely (Bradley, Chapter 5).

Contemporary migration

Demographics changed again when new waves of migration began in the 1970s. This, according to Potter (Chapter 8), marks an increase in the scale of migration and deforestation patterns in Asia and perhaps other parts of the world. From approximately this period onwards, the majority of migrations into tropical forest regions are linked to the need for agricultural land by people who live in extreme poverty. An important factor that contributed to this trend has been natural disasters, resulting in periods of drought. For example, migrants from the Brazilian northeast joined the movements into the Brazilian Amazon region in periods of drought and food deprivation. Historically, natural disasters have been a common trigger of mass movements around the world. The magnitude of this is exemplified by the estimated number of between 25 and 30 million people who have moved to frontier areas since the early 1950s in the Philippines, Indonesia, Vietnam and Thailand (De Konink and Dery 1997, cited in Potter, Chapter 8).

Migration triggered by natural disasters is common to all three tropical continents. Brazil has seen recurrent waves of migrants who moved from the states south of the Amazon basin into the tropical forest region since the 1960s (Browder and Geoffrey 1997). Parallel movements have been seen in countries like Bolivia, Perú and Ecuador (see Perz, Chapter 2). The case of Perú described in much detail by Limachi *et al.* (Chapter 3) reflects a pattern typical for other South American countries with Amazon tropical forest territories. In Perú, migration into the Amazon increased after the 1940s and has gradually intensified since the 1960s and 1970s. Some of the key factors underlying this migration were governments' desire to "reform" economies, the temporary high prices of internationally traded crops, road construction, agricultural development projects, agrarian reform, low interest loans and legislation that promoted agriculture or other economic activities in the tropical forest zone. Push-factors included insurgency, poverty, and high population pressure at home.

The stories told by Tran (Chapter 10) and Acheampong (Chapter 6) are similar to those mentioned above, although timing does not coincide exactly. In Vietnam, spontaneous migration from the northern provinces to the south, that is, the Central Highlands, has increased in the last two decades of the twentieth century, especially since the end of the Vietnam War. The country began economic reform in the mid-1980s and with a heavy focus on the expansion of cash crop production. The cash crop programs were partly expanded into remnant forested areas. This wave coincided with the closing of frontiers and general resource and population pressures in the northern provinces. The high prices achieved for coffee, rubber, sugarcane and pepper between 1991 and 1995 had created strong market incentives, not only for the poor, but also for wealthy entrepreneurs who benefited from these price surges.

Similarly, in Ghana the need for agricultural lands and new livelihood opportunities in the wake of soil degradation in older cocoa production areas account for the continuous advance of the forest frontier. The opening up of forest areas by timber companies explains why, in recent years, the cocoa frontier has moved to the Western part of the country (Acheampong, Chapter 6).

The search for agricultural land by rural poor and the related advancement of the forest frontier are indeed the most typical cases of contemporary tropical forest-related migration. Some important other types, however, include people who escape from violent conflicts, migrating to the forest because they have some hope of carving out a temporary living. This type of migration is probably most prominent in Africa (Bradley, Chapter 5; Price *et al.* 2004) but it also occurred in Latin America and in Asia (de Jong *et al.* forthcoming; Price *et al.* 2004). State movement into forested areas is also ongoing. In addition to clearing forestlands for new plantations and infrastructure projects, transmigration and other such programs as discussed earlier are also often designed to provide poor people from densely populated regions with land and improved welfare. It has been suggested that other factors, such as the desire to expand political control to remote regions of a country, may be behind these state sponsored migration schemes (Potter, Chapter 8; Levang 1997).

The impact of modern migration

As the foregoing discussion suggests, the relationship between migration and deforestation is not a new phenomenon. Different waves of migration made significant contributions to the deforestation process. Despite the larger number of cases discussed in the literature, the precise link between migration to the tropical forest zone and tropical deforestation remains contentious. As Bradley argues in Chapter 5, frequently the understanding of forest-cover change, at least in Africa, remains hidden and is further obscured by false interpretations and assumptions. Other authors in this volume (for example Limachi *et al.*, Chapter 3; Achampong, Chapter 6) observe statistical correlations between people's movements and forest-cover change. Linking other quantitative factors that influence both migration and the impact of migration on the forest, however, is much more difficult. Several of the parameters that shape the impact of migration into tropical forest zones are delimited by Perz (Chapter 2). These parameters include the locale of the settlement and the land-use activities employed. Migrants may move to urban or semi-urban areas. In such cases the impact of migration on forests is present, but certainly much less significant than that caused by migration directly into the forest interiors. Often data used for the analysis of migration trends do not distinguish between urban, semi-urban, rural, or rural–forest destinations (Perz, Chapter 2; Limachi *et al.*, Chapter 3). The dominant land use that migrants employ also has an important influence on how the forest is affected. Contrasting examples are, for instance, crop production versus cattle ranching. The latter has a much larger impact on deforestation rates. Perz (Chapter 2) also shows that government policy at different administrative levels influences which form of land use receives greater promotion and what forms of migration are encouraged.

There is emerging evidence that the phenomenon of migration into tropical forest zones, triggered by poverty, high population density, and scarce agriculture is reaching its final stage, at least in some countries. Potter (Chapter 8) observes this to be the case in countries like Thailand and the Philippines. A more significant trend in these cases appears to be rural out-migration to urban areas. Limachi *et al.* (Chapter 3) suggest this also to be the case in at least one of the examples drawn from Perú. Bradley (Chapter 5) observes this reversal of trends is not only due to the closing of national borders and increased state control of land occupation, but also to the limitation of available land. As a result of which, "new immigrants are not welcome anymore".

Social and cultural changes at the forest frontier

One important demographic change resulting from the phenomenon of migration is that indigenous groups cease to comprise the majority of the population in their own territory (Acheampong, Chapter 6; Tran, Chapter 10). This has been a feature at particular points in history in many locations where outsiders have moved into new lands. Practical results ensue due to the immigrants' dominance of the local economies. Other features of the effects of migration have both an economic and cultural dimension, such as modes of land transfer in the cocoa region of Western Ghana (Acheampong, Chapter 6). Land was previously inherited by a nephew or another male relative of the extended family. Now the same land is passed on to the wife and children thus concentrating ownership within nuclear families. Immigrants into this region did not adopt customary land allocation schemes. As a result of which the renting of land and share cropping appeared. Despite these changes, beneficial to at least some portion of the resident population, local farmers were slow to take up cocoa production. By the time they followed suit, newcomers had cultivated all the suitable cocoa-producing land. This, combined with the conversion of the regional economy into commodity production, led to drastic changes in the commercial value of almost all regional produce. A reduction in local food production and a related increase in the prices of foodstuffs is a significant commonality between many tropical forest regions. Factors other than migration may, however, be the cause of these changes.

Tran's example (Chapter 10) echoes the above discussion. In the Central Highlands of Vietnam, the ethnic composition of the population has changed so that local indigenous groups (locally termed "Ethnic Minorities") are becoming true minorities in the region. The matriarchal inheritance system that previously existed in the region has now been replaced by a patriarchal system. Exacerbated by penetration of state control, the community structure and organization changed profoundly, and customary resource management systems have been eroded. Demand by immigrants has encouraged local residents to rent or sell their land and move deeper into the forest, despite the illegality of such transactions from both a customary and state legal perspective. In recent years these trends have had dramatic impacts. Local people have progressively given up their land and turned to coffee production following the boom in the price of coffee, and are increasingly ceasing their own food production practices. Coffee prices have declined from around US$3 per kg to US$0.27 per kg in recent years, and incomes have declined correspondingly. This has resulted in protests and uprisings in the Central Highlands as people demand the return of their lands (Potter, Chapter 8; Tran, Chapter 10).

Eghenter (Chapter 11) points to additional social changes resulting from itinerant migrants in the forests of East Kalimantan. Here, the immigrants seek a valuable commodity (eaglewood), but have little interest in settling permanently in the region. In order to protect their interests, the local population formulated new customary rules to protect the resources from over exploitation. Some of the new rules involved levying harvesting fees on outsiders seeking to collect eaglewood in their territories. However, because of internal strife, this initiative was seldom enforced. Part of the reason underlying the lack of enforcement was that the new initiatives did not have wide community support. Middlemen tried to establish social ties with some of the local people, for instance by marrying into the communities to obtain kin-only harvesting privileges. Ultimately the eaglewood boom led to social strife and division within villages. In addition to these changes, the trade in eaglewood created a one-product commodity economy, as was the case of the cocoa production zone of West Ghana, resulting in unstable and inflationary market prices for all goods.

Note

1. In more recent years the term "social ecology" has gained prominence, for instance, at the Institute for Social Ecology. At the website of the institute (www.social-ecology.org) social ecology is presented as an integrated scientific analytical framework, a system of values, and an action platform that relates to the concept of political ecology as we have described in a previous publication (Lye *et al.* 2003).

References

Abdoellah, O.S. 1993. *Indonesia Transmigrants and Adaptation: An Ecological Anthropological Perspective.* Berkeley: Center for South Asia Studies.

Adas, M. 1992. From avoidance to confrontation: Peasant protest in pre-colonial and colonial Southeast Asia. In N.B. Dirks, ed., *Colonialism and Culture*, 217–247. Ann Arbor: University of Michigan Press.

Ayuwat, D. 1993. Effects of migration patterns on forest use and forestry projects in a Thai village. *Society and Natural Resources,* 6(2): 195–202.

Bilsborrow, R.E. 2002. Migration, population change, and the rural environment. *Environmental Change and Security Project Report,* 8: 69–94.

Blahna, D.J. 1990. Social bases for resource conflicts in areas of reverse migration. In R.G. Lee, D.R. Field, and W.R. Burch, eds., *Community & Forestry: Continuities in the Sociology of Natural Resources*, Chapter 12. Boulder: Westview Press.

Browder, J.O. and B. Geoffrey. 1997. *Rainforest Cities: Urbanization, Development and Globalization of the Brazilian Amazon.* New York: Columbia University Press.

Brown, R.T. 2002. Society of pediatric psychology presidential address: Toward a social ecology of pediatric psychology. *Journal of Pediatric Psychology*, 27(2): 191–201.

Brunson, M.W. and J.J. Kennedy. 1995. Redefining "multiple-use": Agency responses to changing social values. In R.L. Knight and S.F. Bates, eds., *A New Century for Natural Resources Management*, 143–158. Washington, DC: Island Press.

Bryant, R.L. 1992. Political ecology: An emerging research agenda in Third-World studies. *Political Geography*, 11: 12–36.

Bryant, R.L. and S. Bailey. 1997. *Third World Political Ecology.* London: Routledge.

Charras, M. and M. Pain. 1993. *Spontaneous Settlements in Indonesia: Agricultural Pioneers in Southern Sumatra.* Jakarta: Departement of Transmigration and Bondy, France: ORSTOM.

de Jong, W., D. Rohadi, B. Belcher, R. Mustikasari, and P. Levang. 2003. The political ecology of forest products in Indonesia: A history of changing adversaries. In T.P. Lye, W. de Jong, and K. Abe, eds., *The Political Ecology of Tropical Forests in Southeast Asia: Historical Perspectives*, 107–132. Kyoto: Kyoto Area Studies on Asia, Kyoto University Press and Melbourne: Trans Pacific Press.

de Jong, W., D. Donovan, and K. Abe, eds. forthcoming. *Extreme Conflicts and Tropical Forests.*

Denevan, W.M. 1976. The aboriginal population of Amazonia. In: W.M. Denevan, ed., *The Native Population of the Americas in 1492*, 205–234. Madison: University of Wisconsin Press.

Dove, M.R. 1994. Transition from native forest rubbers to *Hevea Brasiliensis* (Euphorbiaceae) among tribal smallholders in Borneo. *Economic Botany*, 48: 382–396.

Eghenter, C. 1999. Migrants' practical reasonings: The social, political, and environmental determinants of long-distance migrations among the Kayan and Kenyah of the interior of Borneo. *SOJOURN, Journal of Social Issues in Southeast Asia*, 14(1): 1–33.

Ellen, R.F. 1982. *Environment, Subsistence and System: The Ecology of Small-Scale Social Formations.* Cambridge: Cambridge University Press.

Fasbender, K. and S. Erbe. 1990. *Towards a New Home: Indonesia's Managed Mass Migration.* Hamburg: Verlag Weltarchiv GMBH.

Greenberg, J.B. and T.K. Park. 1994. Political ecology. *Journal of Political Ecology*, 1: 1–12.

Ingold, T. 1986. *The Appropriation of Nature: Essays on Human Ecology and Social Relations.* Manchester: Manchester University Press.

Kaimowitz, D., G. Thiele, and P. Pacheco. 1999. The effects of structural adjustment on deforestation and forest degradation in lowland Bolivia. *World Development*, 27(3): 505–520.

Kaskija, L. 2002. *Claiming the Forest. Punan Local Histories and Recent Developments in Bulungan, East Kalimantan.* Bogor: CIFOR.

Kato, T. 1991. When rubber came: The Negeri Sembilan experience. *Southeast Asian Studies*, 29: 109–157.

Lathrap, D. 1970. *The Upper Amazon.* New York : Praeger.

Levang, P. 1997. *La Terre d'en Face—La Transmigration en Indonésie.* Paris: Orstom editions.

Lim Teck Ghee. 1976. *Origins of a Colonial Economy: Land and Agriculture in Perak 1874–1897.* Penang: Universiti Sains Malaysia.

Lohnert, B. and H. Geist. 1999. *Coping with Changing Environments, Social Dimensions of Endangered Ecosystems in the Developing World.* Aldershot: Ashgate.

Lye, T.P., W. de Jong, and K. Abe, eds. 2003. *The Political Ecology of Tropical Forests in Southeast Asia: Historical Perspectives.* Kyoto: Kyoto Area Studies on Asia, Kyoto University Press and Melbourne: Trans Pacific Press.

Lye, T.P. 2004. *Changing Pathways: Forest Degradation and the Batek of Pahang, Malaysia*. Lanham, MD: Lexington Books.

Meggers, B.J. 1971. *Amazonia: Man and Culture in a Counterfeit Paradise*. Chicago: Aldine-Atherton.

Moran, E.F. 2000 (1982). *Human Adaptability: An Introduction to Ecological Anthropology*. 2nd ed. Boulder: Westview.

Padoch, C. 1982. Land use in new and old areas of Iban settlement. *Borneo Research Bulletin*, 14(1): 3–14.

Peet, R., and M. Watts. eds. 1996. *Liberation Ecologies: Environment, Development, Social Movements*. London and New York: Routledge.

Perz, S.G. 2002. Population growth and net migration in the Brazilian Legal Amazon, 1970–1996. In C.H. Wood and R. Porro, eds., *Deforestation and Land Use in the Amazon*, 107–129. Gainesville: University of Florida Press.

Poffenberger, M. ed. 1998. *Stewards of Vietnam's Upland Forests*. Asia Forest Network and Forest Inventory and Planning Institute.

Pokam, J. and W. Sunderlin. 1999. *L'Impact de la Crise Économique sur les Populations, les Migrations et le Couvert Forestier du Sud-Cameroun*. Jakarta: Center for International Forestry Research.

Potter, L. 2003. Forests versus agriculture: Colonial forest services, environmental ideas and the regulation of land-use change in Southeast Asia. In Lye T.P., W. de Jong, and K. Abe, eds., *The Political Ecology of Tropical Forests in Southeast Asia: Historical Perspectives*, 29–71. Kyoto: Kyoto Area Studies on Asia, Kyoto University Press and Melbourne: Trans Pacific Press.

Price, S., W. de Jong, D. Kaimowitz, and D. Donovan. 2004. *Conflict Timber in Subsaharan Africa*. Bogor: Center for International Forestry Research.

Rambo, A.T. 1983. *Conceptual Approaches to Human Ecology*. Research report no. 14. Honolulu: East-West Center.

Rappaport, R. 1967. *Pigs for the Ancestors: Ritual in the Ecology of a New Guinea People*. New Haven: Yale University Press.

Sader, S.A., C. Reining, T. Sever, and C. Soza. 1997. Human migration and agricultural expansion: An impending threat to the Maya Biosphere Reserve. *Journal of Forestry*, 95(12): 27–32.

San Roman, J. 1975. *Perfiles Historicas de la Amazonia Peruana*. Lima: Ediciones Paulinas.

Sellato, B. 1994. *Nomads of the Borneo Rainforest: The Economics, Politics, and Ideology of Settling Down*. Honolulu: University of Hawaii Press.

Sellato, B. 2001. *Forest, Resources and People in Bulungan: Elements for a History of Settlement, Trade and Social Dynamics in Borneo, 1880–2000*. Bogor: Center for International Forestry Research.

Silberbauer, G.B. 1994. A sense of place. In E. Burch and L. Ellanna, eds., *Key Issues in Hunter-Gatherer Research*, 119–143. Oxford: Berg.

Subedi, B.P., C.L. Das, and D. Messerschmidt. 1993. *Trees and Land Tenure in the Eastern Nepal Terai: A Case Study from the Siraha and Saptari Districts, Nepal*. Rome: Community Forestry Case Study No. 9. FAO/SIDA Forests, Trees and People Programme, U.N. Food and Agriculture Organization.

Sunderlin, W.D. and J. Pokam. 2002. Economic crisis and forest cover change in Cameroon: The role of migration, crop diversification, and gender division labor. *Economic Development and Cultural Change*, 50(3): 581–606.

Sunderlin, W.D. and I.A.P. Resosudarmo. 1999. The effect of population and migration on forest cover in Indonesia. *Journal of Environment and Development*, 8(2): 152–169.

Sutlive Jr., V.H. 1978. *The Iban of Sarawak: Chronicle of a Vanishing World*. Prospect Heights: Waveland.

Swasono and Singarimbun. 1986. *Transmigrasi di Indonesia, 1905–1985*. Penerbit Universitas Indonesia.

Swatuk, L.A. forthcoming. Barking up the wrong tree: Tropical forest management and violent conflict in Africa. In D. Donovan, W. de Jong, and K. Abe, eds., *Extreme Conflicts and Tropical Forests*.

Tickle-Degnen, L. and K. Doyle Lyons. 2004. Practitioners' impressions of patients with Parkinson's disease: The social ecology of the expressive mask. *Social Science and Medicine*, 58 (3): 603–614.

Turner, F. 1893. The significance of the frontier in American history. *Report of the American Historical Association*, 199–227 (online) http://www.library.csi.cuny.edu/dept/history/lavender/frontierthesis.html.

Vickery, M. 1998. *Society, Economics, and Politics in Pre-Angkor Cambodia: The 7th–8th centuries*. Tokyo: The Center for East Asian Cultural Studies for UNESCO, The Toyo Bunko.

2

Migrant Characteristics and Land-Use/ Land-Cover Change in the Pan-Amazon Basin: A Comparative Analysis of Brazil, Bolivia, Ecuador and Perú

Stephen G. Perz

Introduction

In recent years, researchers in numerous academic communities have turned their attention to the issues surrounding human occupation and land-cover change in tropical forest regions such as the Amazon basin. Forest loss has many negative biophysical consequences including local soil erosion and runoff into rivers, endemic species loss, loss of environmental functions, and increased carbon emissions (Fearnside 1990; Gash *et al.* 1996). No less significant are the negative social consequences such as conflict over land use, persistent poverty, and poor health outcomes (Hall 1992; Kosinski 1992). As a result, a body of work has emerged that focuses on the "human dimensions" of deforestation (Lambin *et al.* 2001; Liverman *et al.* 1998; Turner *et al.* 1990). This literature features population as a factor underlying land-cover change. However, the empirical findings linking population size or growth to impact on land-cover change are mixed (Geist and Lambin 2002; Kaimowitz and Angelsen 1998; Lambin *et al.* 2001). One reason for this is that aggregate population change itself obscures distinct demographic processes that contribute in various ways to the size, growth and distribution of populations (Shryock and Seigel *et al.* 1976). Among these demographic processes, migration has frequently been singled out as a cause of environmental problems, particularly in regions such as the Amazon. However, the specific environmental consequences of migration are more often assumed than explicated, and the impact of migration in frontier regions such as the Amazon can be difficult to measure (Perz 2002a).

This chapter considers the linkages between migration and land-cover change in the Pan-Amazon basin, by which I mean certain states of Brazil,

Bolivia, Perú and Ecuador.[1] First, I review theoretical perspectives on the correlations between population and migration, and land-use and land-cover change. Second, I present data on land-cover change, migration, and land use in the Pan-Amazon basin to enable a comparative analysis of the relationship between migration and land-cover change. The findings suggest a picture that is considerably more complicated and nuanced than generally believed, motivating discussion of other factors involved that may mediate the migration–environment link in the Pan-Amazon. I will focus on the intervening factor of land use, and also provide a contextualized analysis of political-economic differences among the countries under consideration.

Theories of migration and the environment: The case of land-cover change in the Amazon

Neo-Malthusian and Boserupian notions feature prominently in population–environment interaction theory. The work of Malthus (1989[1798]) is among the earliest drawn upon in this regard. He argued that population growth leads to agricultural expansion and ultimately to land degradation and famine. Criticisms of Malthusian perspectives gave rise to Boserup's (1965) argument that population growth leads to sustainable land use via intensification due to technological changes. This in turn prompted the emergence of "induced innovation" literature that emphasized the role of market expansion as a driver of agricultural technology adoption during the "Green Revolution" in the 1960s (Binswanger and Ruttan 1978). Induced innovation theory fueled debates over the relative importance of population and markets, and historical, cultural and political factors for land use and environmental degradation. From these discussions, "induced intensification" theory developed a broader framework for understanding both subsistence- and market-oriented land-use practices, taking into account factors that mediate the impact of population on the environment, such as state policies, land tenure, and the biophysical environments themselves (Turner *et al.* 1993). Since the 1980s, numerous publications have identified other mediating factors concerning the population–environment link (Arizpe *et al.* 1994; Mazur 1994; Ness *et al.* 1993; Panayotou 1996; Pebley 1998; Torres and Costa 2000). New theoretical works further argue that the effect of population and migration on environments depends on the scale or level of observation (Gibson *et al.* 2000; Wood 2002).

Researchers linking population to deforestation have encountered similar difficulties, namely that the relationships involved are indirect and instead are mediated by many other factors (Geist and Lambin 2002; Kaimowitz and Angelsen 1998; Lambin *et al.* 2001). This is the case in the Pan-Amazon basin (Perz 2002a; Pfaff 1999; Pichón 1997; Rudel and Horowitz 1993; Wood 1992, 2002; Wood and Skole 1998).

At the same time, researchers have paid increasing attention to the specific role of migration as a potential means of rapid population redistribution that may generate environmental degradation. Bilsborrow (1987) applies Davis' (1963) theory of demographic-economic responses to migration. In regions with high population densities, land scarcity or land conflicts, and technological or market bottlenecks that limit increases in agricultural productivity, out-migration serves as a rational response for households. Migrants then move either to growing cities or to rural environments with lower population densities. Migration may also be motivated by the inability of local ecosystems to support large human populations. Population analysts have also recognized that migration is a potentially important demographic mechanism with contingent environmental ramifications (Hogan 1991; Wood 1992; Perz 2002a).

During the past few decades, the Pan-Amazon basin has experienced population growth driven by high fertility rates, declining mortality, and rapid in-migration. In many cases, the latter of these processes has been particularly intense, spurring new land settlement, agricultural expansion, and deforestation. However, the prototypical scenario of in-migration followed by agricultural activities and deforestation is not the only possible course of events (Wood 1992; Perz 2002a). One alternative, involving largely urban settlement, would not directly lead to deforestation (though it may by indirect means, for example by generating demand for local agricultural products; see Llanque, Chapter 4). An alternative possibility is that in-migration ceases but agricultural expansion and deforestation continue. This may occur, for example, during periods of economic growth in consolidated areas that feature forest on properties held by migrants from past years. Many observers have noted that migration to the Brazilian Amazon slowed down after the mid-1980s (Martine 1994; Ozorio de Almeida and Campari 1995; Browder and Godfrey 1997; Perz 2002a; INPE 2002). These alternative scenarios indicate that "rural in-migration-deforestation" is not the only possible outcome for the Pan-Amazon region, however, data from satellite imagery indicate continued forest loss throughout the 1990s.

As the above discussion indicates, the impact of migration on land-cover change is likely to be contingent on a variety of factors. Three types of potential factors can be identified: 1) the characteristics of the migrants themselves; 2) land-use practices of migrants and others; and 3) the broader political-economic context in which migrants are situated.

Migrant characteristics are an issue that arises out of the urban migration scenario and features prominently in many migration theories that highlight various forms of migrant selectivity (Goldscheider 1983; Massey *et al.* 1993; Ritchey 1976; Wood 1982). People often move in order to seek improved training and job opportunities in urban labor markets, particularly during the transitional stage of early adulthood.

Migrant land use is a significant concern in largely agricultural regions such as the Pan-Amazon where migrants may adopt a variety of farming systems. This issue is related to migrant selectivity; migrant land use depends on cultural background, human, financial and social capital, and access to land (Ozorio de Almeida and Campari 1995; Perz 2002a; Pichon 1997; Schmink and Wood 1992; Serrao and Homma 1993; Walker *et al.* 2002). For example, if one thousand well-capitalized migrants establish large-scale ranching operations the land-cover impact is likely to be much higher than if one thousand poor migrants establish small-scale farms in the same locale.

Finally, the political-economic context of migration to environments such as the Pan-Amazon basin may moderate land-cover impacts. In Brazil during the 1970s and 1980s, many state policies actively encouraged frontier migration and rapid land-cover conversion (Binswanger 1991), while policies in other Amazonian countries had somewhat more limited effects (Rudel and Horowitz 1993; Pacheco 1998). Broader national circumstances may condition the size and composition of migratory flows to frontier regions, thereby indirectly influencing the extent of the resulting deforestation.

However, quantitative studies of migration and land-cover change, in the Amazon region or elsewhere, remain rare (Perz 2002a). Scarcer still are comparative analyses of the migration–land-cover change linkage for multiple countries (but note Bilsborrow 2002; Perz *et al.* 2002). The sections that follow present recent data on land-cover change, migration and land use in the Amazonian states of Brazil, Bolivia, Ecuador, and Perú. These data enable comparative analysis of the relative significance of migration and other factors on deforestation in the basin. Emphasis is placed here on the relevance of the size and composition of migratory flows in order to address questions about migrant selectivity and land-cover change.

Land-cover change in the Pan-Amazon

Table 2.1 presents estimates of forest loss for the states of Brazil's Legal Amazon from 1978 to 2000. This is a government planning region that

Table 2.1. Deforestation in the Brazilian Amazon, 1978–2000

State	Land area deforested[a] (%)			Average annual deforested[b] (%)		Total land area (km²)
	1978	1990	2000	1978–1990	1990–2000	
Classical Amazon						
Acre	1.63	6.70	10.26	0.42	0.30	153,698
Amapá	0.14	0.91	1.38	0.06	0.04	142,359
Amazonas	0.11	1.42	1.93	0.11	0.04	1,567,954
Pará	4.52	11.57	16.05	0.59	0.37	1,246,833
Rondônia	1.76	14.05	24.39	1.02	0.86	238,379
Roraima	0.04	1.69	2.84	0.14	0.10	225,017
Classical Amazon	1.82	6.02	8.75	0.35	0.23	3,574,239
Other Amazon						
Maranhão	19.39	28.34	31.64	0.75	0.27	329,556
Mato Grosso	2.22	9.27	15.97	0.59	0.56	901,421
Tocantins	1.15	8.26	9.68	0.59	0.12	277,322
Other Amazon	5.77	13.25	18.23	0.62	0.42	1,508,298
Legal Amazon	2.99	8.17	11.56	0.43	0.28	5,082,537
Total area deforested (km²)	152,200	415,200	587,727	21,917	17,253	

Sources: Deforestation: INPE (2002) analysis of Landsat MSS and TM images; land area: IBGE (1991a: 169).

Notes

a: Percent deforested refers to deforested land area as a percentage of total land area, as of the year stated.

b: Average annual deforestation refers to net forest loss per year as a percentage of total land area.

encompasses nine states and approximately 5 million km^2. The "Classical" Amazon encompasses the northernmost states containing more recent settlement; the "Other" Amazon comprises states on the southern and eastern fringes of the basin housing older settlements. Deforestation estimates in this table are based on analyses of Landsat MSS and TM satellite imagery (INPE 2002). Deforested area as a percentage of total land area rose from 3 percent in 1978 to nearly 12 percent in 2000. Within Brazil's Legal Amazon, deforested land area varies substantially, from pre-frontier states such as Amazonas where it is only 2 percent to frontier states like Rondônia, where it is 24 percent and rising rapidly, to old frontiers such as Maranhão, where it is over 30 percent and rising slowly. Average annual deforestation has changed over time, with a slight decline from the 1980s at 21,900 km^2 per year, or 0.43 percent, to 17,200 km^2 per year during the 1990s, or 0.28 percent.

Table 2.2 allows for comparative analysis by presenting available deforestation estimates for states in the three Andean Amazon countries. In Bolivia, I consider only those departments entirely in the eastern lowlands: Beni, Pando, and Santa Cruz—an area of 485,000 km^2 (Pacheco 1998: 57). The estimates presented are drawn from analyses of Landsat TM images for the lowlands for 1985 and 1990 (CUMAT 1992, cited in Pacheco 1998: 57). During this time, 460 km^2 were deforested per year, or 0.09 percent of the region. Approximately 15,300 km^2 (or 3.2 percent of the region) was deforested as of 1985, rising to 17,700 (or 3.6 percent) in 1990. Deforestation rates and percentages were highest in Santa Cruz and lower in remote areas of the lowlands.

The Ecuadorian Oriente, or Amazon lowlands, includes five provinces—Morona-Santiago, Napo, Pastaza, Sucumbios and Zamora-Chinchipe—which encompass 130,000 km^2 (INEC 1992: 48). According to one recent Landsat-based estimate, about 15 percent of the Oriente land was deforested as of 1996 (Rodriguez 2001). Deforestation is relatively high in Morona-Santiago, Napo, and Sucumbíos. The lack of comparable data prevents calculation of deforestation rates in the Oriente, but Rudel and Horowitz (1993: 44) note a very high national deforestation rate of 2.3 percent per year in the 1977–1985 period.

The Peruvian selva includes both high and low lying forests. These areas cut across department boundaries and as only department-level deforestation estimates are available I consider the selva as the departments of Amazonas, Loreto, Madre de Dios, San Martín and Ucayali, an area of 650,000 km^2 (INEI 1994: 48; Limachi *et al.*, Chapter 3).[2] Perú has conducted forest inventories of the selva (INRENA nd, cited in INEI 1997: 283). Overall, deforestation rose from 36,000 km^2 or 5.5 percent in 1985 to 45,000 km^2 or 7.0 percent in 1990, implying an annual average rate of nearly 2,000 km^2 or 0.29 percent. However, state-level estimates indicate large disparities in the extent of deforestation among departments, with Amazonas and San Martín showing 25 percent deforestation or more by 1990, with average annual rates at around one percent.

The disparate data sources in Tables 2.1 and 2.2 render comparative analysis prone to misleading interpretations. They do suggest, however, contrasting patterns if we focus on interpretations of Landsat imagery. Brazil has a relatively high percentage of deforested land (about 8 percent in 1990) and a relatively high deforestation rate (about 0.4 percent per year),

Table 2.2. Deforestation in the Bolivian Lowlands, Ecuadorian Oriente, and Peruvian Selva, 1980s–1990s

Country, Unit	**Land area deforested[a] (%) Time 1**	**Time 2**	**Average annual deforested[b] (%)**	**Land area (km²)**
Bolivian Lowlands	**1985**	**1990**	**1985–90**	
Beni	0.8	1.1	0.05	196,270
Pando	2.5	2.9	0.07	63,830
Santa Cruz	5.4	6.1	0.14	224,690
Overall percent	3.2	3.6	0.09	484,790
Total area deforested (km²)	15,370	17,670	460	
Ecuadorian Oriente		**1996**		
Morona-Santiago	nd[c]	25.0	nd	24,606
Napo	nd	16.0	nd	37,682
Pastanza	nd	5.0	nd	29,137
Sucumbíos	nd	17.0	nd	22,981
Zamora-Chinchipe	nd	13.0	nd	16,014
Overall percent	nd	15.0	nd	130,420
Total area deforested (km²)	nd	19,626	nd	
Peruvian Selva	**1985**	**1990**	**1985–90**	
Beni	0.8	1.1	0.05	196,270
Amazonas	33.0	37.8	0.96	39,249
Loreto	2.1	2.8	0.15	368,852
Madre de Dios	0.5	0.9	0.08	85,183
San Martín	20.8	26.4	1.12	51,253
Ucayali	4.1	5.6	0.30	102,411
Overall percent	5.5	7.0	0.29	646,948
Total area deforested (km²)	35,844	45,240	1,879	

Sources: Bolivia: CUMAT (1992) analysis of Landsat images, in Pacheco (1998: 57); land area: INE (1997a: 5). Ecuador: Rodriguez (2001); land area: INEC (1992: 48). Perú: INRENA (nd) forest inventory, cited in INEI (1997: 283); land area: INEI (1994: 48).

Notes

a: Percent deforested refers to deforested land area as a percentage of total land area, as of the year stated.

b: Average annual deforestation refers to net forest loss per year as a percentage of total land area.

c: No data available.

Photo 2.1. Colonization landscape in the eastern Brazilian Amazon

and Bolivia has lower percentages deforested (about 4 percent in 1990) and a modest deforestation rate (<0.1 percent per year). Furthermore, there is substantial variation among states within and among countries. These contrasts raise questions about the extent to which patterns of demographic change correspond to deforestation estimates.

Migration in the Pan-Amazon basin

This section presents a comparative analysis of migration among the Amazonian regions of Brazil, Bolivia, Ecuador and Perú. I first focus on the absolute and relative scales of migration into the Pan-Amazon in order to determine whether in-migration has proved more intense in areas experiencing faster forest loss. The analysis will later turn to the issue of migrant composition in an effort to better understand the characteristics of recent migrants to the basin, and to further establish whether those characteristics afford a clearer view of any linkage between demographic processes and land-cover change.

Definitions and measures of migration

Before discussing the data and proceeding toward an interpretation, some comments about migration measurement are in order. These measurements can be problematic, particularly in a frontier region of enormous size such as the Pan-Amazon basin. There are diverse methods for estimating migration, and each has its shortcomings. In this chapter, I rely on direct estimation of in-migration to the Pan-Amazon, in order to disaggregate migrants into subpopulations defined in terms of various characteristics such as previous residence, rural or urban settlement, age, and socioeconomic status. For present purposes, I define migrants as persons who moved between states (or departments as in Bolivia and Perú, or provinces as in Ecuador) during the five year period prior to a given census. Given available data, this definition allows comparability of migration rates across the countries that share the Pan-Amazon basin. While measurement of migration based on a question about residence five years before certainly misses some people who made significant geographical moves, this is regarded as the single best question for migration analysis (Shryock and Seigel *et al.* 1976; UN 1970).

Migration estimates in the Pan-Amazon basin

Table 2.3 presents estimates of inter-state migrants over five-year intervals as a percentage of total populations enumerated in various censuses in the Pan-Amazon. Column one shows the population age ≥5 who declared a residence five years previously. This serves as the population subject to migration over a five-year period. Columns two and three provide counts and percentages of migrants out of the ≥5 age groups who declared a residence five years prior. For purposes of the present analysis, I regard all other persons as non-migrants, even if they settled in the Pan-Amazon at some earlier point in time. Column three thus provides percentages of the population in the Pan-Amazon who were inter-state migrants in the previous five-years as shown by the censuses of the early 1990s in the countries under consideration.

Migrant counts and percentages vary substantially across the Pan-Amazon basin. The largest figures are for the Brazilian Legal Amazon, which shows about 850,000 in-migrants during 1986–1991, primarily in the "Classical" Amazon states. The 1991 census also indicates that pre-frontier states such as Amazonas had exhibited low percentages (about 3 percent) of migrants since 1986, while frontier states such as Rondônia and Mato Grosso have double-digit percentages, as does Roraima. However, other frontier states such as Pará have low migrant percentages. This may reflect the slowdown in migration rates noted by other analysts (Martine 1994; Ozorio de Almeida and Campari 1995; Perz 2002a). One can view in-migration rates in terms of population flows, and use these "flow indicators" to determine whether they correspond to rates of land-cover conversion. In the Brazilian Legal Amazon, migration figures in Table 2.3 correspond to data on rates of forest loss presented in Table 2.1, as both are high for the 1980s in Rondônia and low in Amazonas. However, Mato Grosso exhibits a high rate of in-migration and only a moderate rate of forest loss for this period, Pará shows a low rate of in-migration and a moderate rate of deforestation, and Roraima exhibits a high rate of in-migration but a low rate of forest loss.

Table 2.3 Inter-state migrants as a percentage of population (≥5), Amazonian Brazil, Bolivia, Ecuador, and Perú (late 1980s–early 1990s)

Country, Administrative Unit	**Population age ≥5**[a]	**Migrants, last 5 years**	**Migrants, last 5 years** (%)
Brazilian Legal Amazon	**1991**	**1986–91**	**1986–91**
Classical Amazon			
Acre	354,399	12,846	3.6
Amapá	241,760	23,261	9.6
Amazonas	1,787,413	56,234	3.1
Pará	4,218,105	202,422	4.8
Rondônia	970,266	121,580	12.5
Roraima	185,044	33,712	18.2
Total Classical Amazon	7,756,987	450,055	5.8
Other Amazon			
Maranhão	4,182,888	97,712	2.3
Mato Grosso	1,762,991	222,887	12.6
Tocantins	787,059	77,708	9.9
Total Other Amazon	6,732,938	398,307	5.9
Total Brazilian Legal Amazon	14,489,925	848,362	5.9
Bolivian Lowlands	**1992**	**1987–92**	**1987–92**
Beni	225,343	19,344	8.6
Pando	31,094	4,897	15.7
Santa Cruz	1,130,140	88,181	7.8
Total Bolivian Lowlands	1,386,577	112,422	8.1
Ecuadorian Oriente	**1990**	**1985–90**	**1985–90**
Morona-Santiago	69,313	7,074	10.2
Napo	85,102	13,824	16.2
Pastanza	35,367	6,787	19.2
Sucumbíos	64,026	16,684	26.1
Zamore Chinchipe	54,935	9,287	16.9
Total Ecuadorian Oriente	308,743	53,656	17.4
Peruvian Selva	**1993**	**1988–93**	**1988–93**
Amazonas	275,407	22,388	8.1
Loreto	551,919	29,743	5.4
Madre de Dios	55,396	12,577	22.7
San Martín	454,008	69,908	15.4
Ucayali	259,196	43,596	16.8
Total Peruvian Selva	1,595,926	178,212	11.2

Sources: Brazil: IBGE (1996); Bolivia: INE (1992: Table 4); Ecuador: INEC (1992: Table 4); Perú: INEI (1994: Table 5).

a: The population observed is the population age ≥5 that declared a place of residence 5 years prior to the census. All figures for migrants and non-migrants in this table refer to persons age ≥5.

In the Bolivian lowlands, while migrant counts based on 1992 census data are relatively low for the 1987–1992 period, migration percentages are somewhat higher, perhaps indicative of more recent frontier expansion there than in Brazil. Whereas the population of the Brazilian Legal Amazon consisted of about 6 percent migrants during the previous five years, in the Bolivian lowlands, this figure is 8 percent. The correspondence of in-migration rates with rates of land-cover conversion (shown in Table 2.2 for Bolivia) is also moderate in Bolivia. Pando shows the highest rate of in-migration but a moderate rate of forest loss, while Santa Cruz exhibits a moderate rate of in-migration and a high rate of forest loss. Further, rates of in-migration are lower in the Bolivian lowlands than in the Brazilian Legal Amazon, while rates of forest loss are higher in Brazil than Bolivia.

While counts in the Ecuadorian Oriente are low, nowhere are migration rates in the late 1980s as high as in the Oriente, with an overall rate of over 17 percent. The 1990 census data indicate heavy in-migration since 1985, with migrants comprising 20 percent or more of the ≥5 age groups in Pastaza and Sucumbíos. The lack of data on rates of land-cover change over time in the Oriente leaves open questions about the correspondence with rates of in-migration. Available data on the stock of deforested land in the Oriente as of 1996 shows little correspondence with indicators of migration flows there.

Departments of the Peruvian selva show substantial numbers of migrants during 1988–1993, and migration rates are also high by Pan-Amazonian standards for that period, with 11 percent of the 1993 selva population having migrated in the previous five years. In-migration rates are over 15 percent in Madre de Dios, San Martín, and Ucayali. However, the correlation with deforestation rates is again limited. On the one hand, San Martín shows a high percentage of in-migrants and a high rate of forest loss, and Loreto shows low rates for both. Further, a comparison of Perú and Bolivia indicates higher rates of in-migration and forest loss in the Peruvian selva than in the Bolivian lowlands. On the other hand, Madre de Dios and Ucayali both display high in-migration rates and low deforestation rates. And, if we compare the Peruvian and Brazilian situations, we find higher in-migration rates in Perú than Brazil and yet higher rates of forest loss in Brazil than Perú.

Migrant composition

These findings broadly agree with those of other comparative analyses of overall migration and land-cover change in the Pan-Amazon region (Perz

2002a; Perz *et al.* 2002). There appears to be a limited correspondence between migration and land-cover change, at least at the level of states (and equivalent administrative units). This then begs the question as to whether certain types of migrants may have greater impacts on land-cover change than others. These questions require us to look beyond aggregate migrant flows to consider migrant composition.

Rural and urban migration

As noted earlier, the stereotypical scenario for migration to the Amazon involves rural land settlement followed by deforestation. The region's urbanization, however, implies that most migrants may instead be settling in urban areas, generating (at most) indirect impacts on land-cover change.

Available data allow for rural/urban disaggregation of migrants in Brazil, Bolivia and Ecuador for census enumerations in the early 1990s. Table 2.4 presents rural and urban migration figures for states in these three Pan-Amazonian countries. Note that migration estimates here may not correspond with those provided in Table 2.3 since some migrants did not clearly declare whether they were in a rural or urban area. The figures in Table 2.4 indicate the same basic pattern in virtually all states of all three countries. During the late 1980s and early 1990s in the Brazilian Legal Amazon, the Bolivian lowlands, and the Ecuadorian Oriente, most migrants settled in urban areas, and migration percentages are higher in urban areas. The rightmost column indicates the percentage of migrants living in urban areas, and this varies from about 65 percent in Brazil to nearly 80 percent in Bolivia and Ecuador. Clearly, by the late 1980s, the Pan-Amazon was largely an urban frontier (Becker 1995; Browder and Godfrey 1997).

Rural and urban settlement patterns also show a fairly strong correspondence with forest loss in the Pan-Amazon. This is most evident in the Brazilian Legal Amazon. There, two states with the highest population percentages of rural in-migrants during the 1986–1991 period—Rondônia and Mato Grosso—also exhibit some of the highest deforestation rates, especially moving into the 1990s, when migrants' impacts on land cover would become apparent. However, the rural in-migration rate was still fairly low in Pará, which shows a relatively high rate of forest loss, and rural in-migration was high in Roraima, which shows a low rate of forest loss. In the Bolivian lowlands this correspondence is much less clear, for the percentage of rural in-migrants is highest in Pando, but the rate of forest loss is highest in Santa Cruz. And in the Ecuadorian Oriente, the lack of multi-temporal land cover data prevent direct comparisons of migrant percentages and deforestation rates over time.

Table 2.4. Rural and urban migration to the Brazilian Legal Amazon, Bolivian Lowlands, and Ecuadorian Oriente, late 1980s–early 1990s

	Rural areas			Urban areas			
Country, State	Population age ≥5[a]	Migrants, last 5 years[b]	Migrants, last 5 years[b] (%)	Population age ≥5[a]	Migrants, last 5 years[b]	Migrants, last 5 years[b] (%)	Migrants in urban areas (%)
Brazilian Legal Amazon							
Classical Amazon							
Acre	131,089	3,462	2.6	223,310	9,384	4.2	73.0
Amapá	44,005	3,687	8.4	197,755	19,574	9.9	84.1
Amazonas	490,943	2,949	0.6	1,296,470	53,285	4.1	94.8
Pará	1,962,744	96,956	4.9	2,255,361	105,466	4.7	52.1
Rondônia	403,448	52,035	12.9	566,818	69,545	12.3	57.2
Roraima	65,480	8,135	12.4	119,564	25,577	21.4	75.9
Total Classical Amazon	3,097,709	167,224	5.4	4,659,278	282,831	6.1	62.8
Other Amazon							
Maranhão	2,471,562	40,200	1.6	1,711,326	57,512	3.4	58.9
Mato Grosso	465,053	68,094	14.6	1,297,938	154,793	11.9	69.4
Tocantins	325,977	22,200	6.8	461,082	55,508	12.0	71.4
Total Other Amazon	3,262,592	130,494	4.0	3,470,346	267,813	7.7	67.2
Total Brazilian Legal Amazon	6,360,301	297,718	4.7	8,129,624	550,644	6.8	64.9
Bolivian Lowlands							
Beni	73,403	4,281	5.8	151,846	15,078	9.9	77.9
Pando	22,557	3,152	14.0	8,540	1,753	20.5	35.7
Santa Cruz	306,156	17,015	5.6	823,949	71,681	8.7	80.8
Total Bolivian Lowlands	402,116	24,448	6.1	984,335	88,512	9.0	78.4
Ecuadorian Oriente							
Morona-Santiago	39,288	1,700	4.3	29,929	5,278	17.6	75.6
Napo	40,770	2,774	6.8	44,063	10,751	24.4	79.5
Pastanza	17,850	1,162	6.5	17,422	5,530	31.7	82.6
Sucumbíos	23,191	3,002	12.9	39,781	12,628	31.7	80.8
Zamora-Chinchipe	28,371	2,266	8.0	26,520	6,977	26.3	75.5
Total Ecuadorian Oriente	149,470	10,904	7.3	157,715	41,164	26.1	79.1

Sources: Brazil: IBGE (1996); Bolivia: INE (1992: Table 4); Ecuador: INEC (1992: Table 4).

Notes

a: The population observed is the population age ≥5 that declared a place of residence 5 years prior to the census. All figures in this table refer to persons age ≥5.

b: Migration estimates here may not total those given in Table 3 since some migrants did not clearly declare whether they were in a rural or urban area.

Migrant socioeconomic status

This section considers the socioeconomic status of migrants in order to ascertain whether predominance of young adults among migrants is reflected in socioeconomic differences between recent migrants and the rest of the population in the Pan-Amazon. For present purposes, given data limitations, I define socioeconomic status in terms of literacy and the industry sector of one's primary occupation. This section focuses on migrants in the Brazilian Legal Amazon and the Bolivian lowlands, for which census data allow comparison.

Figures 2.1 through 2.4 afford this analysis by examining the economically active population, defined here as persons age ≥7 in Bolivia and age ≥10 in Brazil who declare an economic activity. Figures 2.1 and 2.3 compare the distribution of recent migrants and other persons across industry sectors, ranging from agriculture to finance, for the Brazilian Legal Amazon and Bolivian lowlands, respectively. In both regions of the Pan-Amazon, economically active migrants tend to be underrepresented in agriculture, predominating instead in sectors such as manufacturing, commerce, and miscellaneous service industries.

Figures 2.2 and 2.4 make similar comparisons for each state of the Brazilian Legal Amazon and the Bolivian lowlands respectively. To the left is the overall percentage of migrants in the economically active population for a given state or department. If one uses this point as a benchmark, one can see for other points along a given line which sectors have proportionally more migrants than overall. The heaviest lines represent the regional averages for a given country. Patterns vary among the states and departments, and in ways relevant to understanding migration and deforestation. In Figure

Photo 2.2. Small farm colonist in the eastern Brazilian Amazon

Figure 2.1. Labor-force composition, inter-state migrants in 1986–1991 and other persons, Brazilian Amazon, 1991

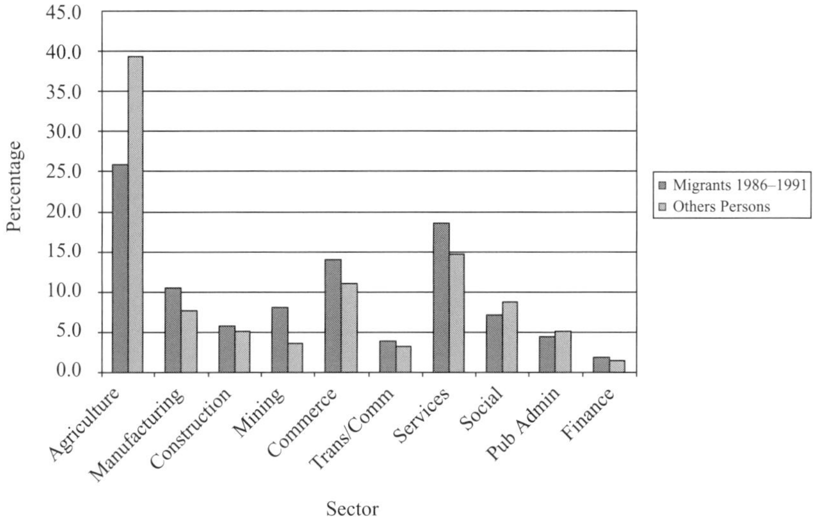

Figure 2.2. Migrants in 1986–1991 as a percentage of the economically active population, by sector, states of the Brazilian Amazon, 1991

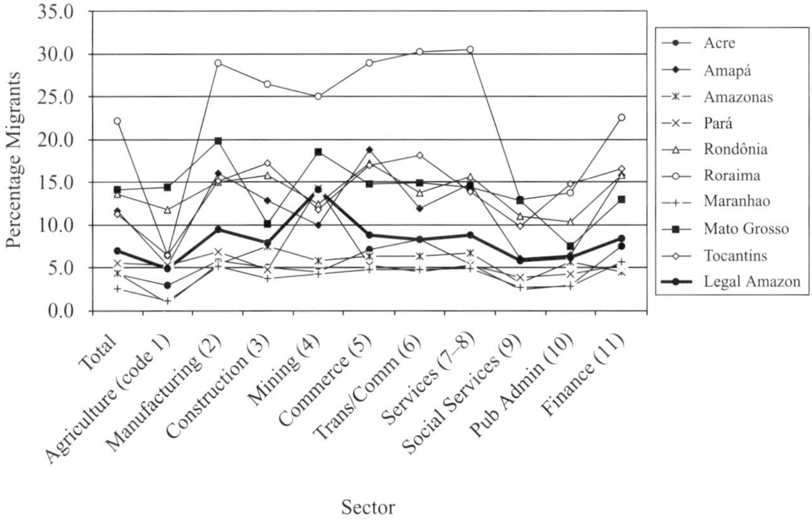

Figure 2.3. Labor-force composition, inter-department migrants in 1987–1992 and other persons, Bolivian Lowlands, 1992

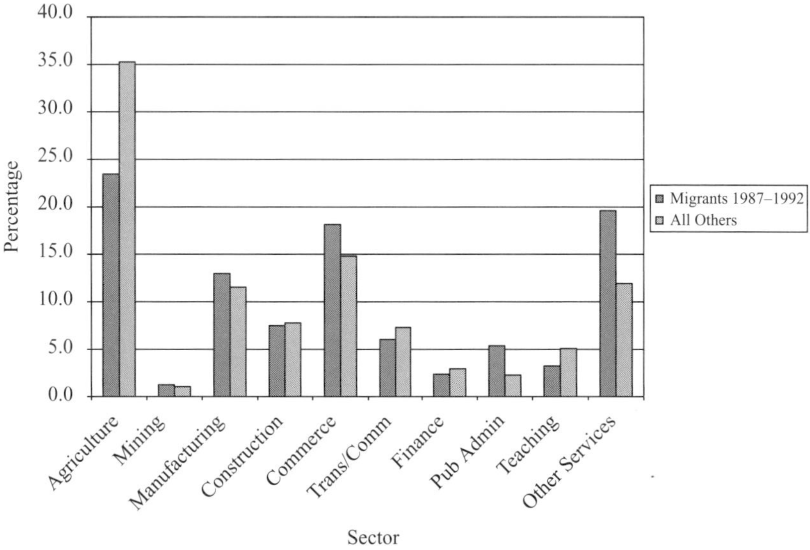

Figure 2.4. Migrants in 1987–1992 as a percentage of the economically active population, by sector, departments of the Bolivian Lowlands, 1992

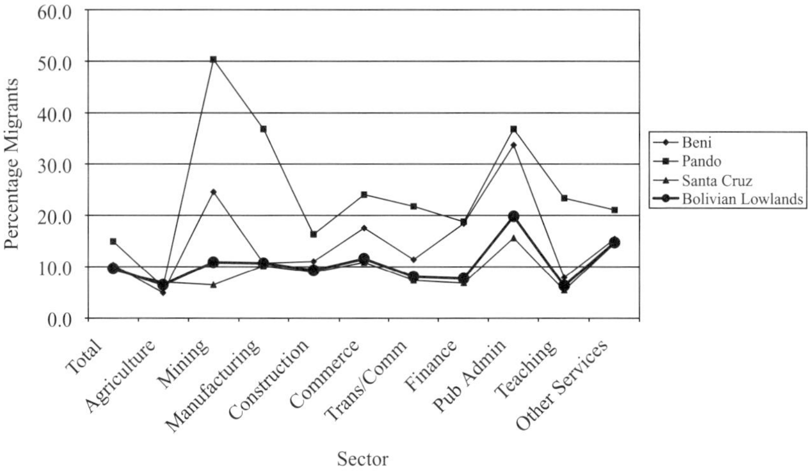

2.2, we find an explanation as to why migration rates to Roraima were so high and yet forest loss so low. Only about 5 percent of people working in agriculture in that state were migrants, while migrants constituted 25–30 percent of the labor-force in manufacturing, construction, mining, commerce, transportation/communications, and miscellaneous service industries. In contrast, in Rondônia and Mato Grosso, migrants are well-represented in agriculture relative to other industry sectors, a finding consistent with high rural percentages of migrants, high overall migration, and substantial rates of forest loss in those states, especially during the 1990s. Similarly, we observe an explanation as to why migrant percentages were so high in Pando in Bolivia, while rates of forest loss there were so low. The same reason that explains the pattern in Roraima is apparent here: migrants are underrepresented in agriculture, instead opting to move into other sectors such as mining, manufacturing, and public administration. Alternatively, the distribution of migrants in Santa Cruz is very even among industries, with migrants more or less equally represented in agriculture as well as commerce and other sectors. This helps explain why migration is still relevant to the fact that the rate of forest loss is higher in Santa Cruz than in Pando.

Intervening factors: Land use

We have observed that the composition of migrants among rural/urban areas and industry sectors helps to explain some perplexing findings about population, aggregate migration and forest loss. Nonetheless, it is important to recognize that migration in and of itself is by no means a complete explanation for land-cover change in the Amazon, if for no other reason than because of the processes that occur between in-migration and forest loss. No matter how many migrants arrive in a given area, and no matter what their characteristics are—even if they all settle in rural areas and practice agriculture—forest outcomes can still vary, sometimes substantially. This brings us back to the intervening factors that moderate the effect of migration on land-cover change. A key example of such an intervening factor between demographic processes and environmental outcomes is land use (Turner *et al.* 1990). Hence, migration must be viewed as an indirect determinant of land-cover change, with land-use practices intervening and moderating this relationship (Perz 2002a). Local populations—whether long-term residents or recent migrants—differ in their access to capital and land. This reflects and affects their land use, which in turn has direct consequences for land-cover change. Two indicators arguably reflect land-use practices in the

Photo 2.3. Pasture in the eastern Brazilian Amazon

Pan-Amazon basin to a considerable extent, namely the land area planted with annual and perennial crops, and the cattle density (Serrão and Homma 1993; Pichón 1996).

Table 2.5 presents the percentage of total land area under cultivation and the head of cattle per km^2 for the Pan-Amazon basin in Brazil, Bolivia, Ecuador and Perú, based on enumerations from agricultural censuses and other state data sources. In general, the percentage of land deforested reflects the percentage of cultivated land; both are higher in Rondônia and the "Other" Amazon states. However, the percentage of planted land declined from 1985 to 1996 nearly everywhere in the Brazilian Amazon, while deforestation increased. Cattle density accounts for this discrepancy. Not only is deforestation greater where cattle are more prevalent, but cattle density increased at a particularly rapid rate where deforestation most increased in Rondônia, Pará, and the "Other" Amazon states. The importance of "pecuarização" or the change from a forest extractive to an agriculture based subsistence in the Brazilian Amazon (Perz 2002b) and the limited labor required for cattle farming (Serrão and Homma 1993; Pichón 1996) helps account for the limited correlation between population growth, net migration and deforestation.

In the Bolivian lowlands, recent data indicates expanding crop cultivation and cattle farming. While the land-use figures presented here refer to a period somewhat later than available deforestation estimates, national figures show similar trends from 1985. The expansion of land cultivation and cattle farming is most rapid in Santa Cruz, precisely where deforestation is occurring at the greatest rate.

In the Ecuadorian Oriente, all states show increases in both the percentage of land cultivated and cattle density. If an earlier estimate of the national deforestation rate is to be believed (Rudel and Horowitz 1993: 44), then the

Table 2.5. Indicators of land cultivation and cattle ranching in Amazonian Brazil, Bolivia, Ecuador and Perú, 1980s–1990s

State	Land area cultivated[a] (%)		Cattle/km²		Total land area[a]
	1980s	1990s	1980s	1990s	
Brazilian Legal Amazon	**1985**	**1996**	**1985**	**1996**	
Classical Amazon					
Acre	0.4	0.5	2.2	5.5	153,698
Amapá	0.2	0.1	0.3	0.4	142,359
Amazonas	0.2	0.2	0.3	0.5	1,567,954
Pará	0.9	0.6	2.8	4.9	1,246,833
Rondônia	2.2	1.8	3.2	16.5	238,379
Roraima	0.1	0.6	1.4	1.9	225,017
Total Classical Amazon	0.6	0.5	1.5	3.4	3,574,239
Other Amazon	2.6	2.7	8.7	15.6	1,508,298
Maranhão	3.2	2.5	9.0	11.8	329,556
Mato Grosso	2.4	3.2	7.3	15.9	901,421
Tocantins	2.4	1.0	13.0	18.8	277,322
Total Other Amazon	2.6	2.7	8.7	15.6	1,508,298
Total Brazilian Legal Amazon	1.2	1.1	3.6	7.0	5,082,537
Total Land/Cattle[b]	58,792	57,265	18,485,510	35,538,831	
Bolivian Lowlands	**1988**	**1999**	**1990**	**1999**	
Beni	0.1	0.1	11.2	14.1	213,564
Pando	0.2	0.2	0.2	0.3	63,827
Santa Cruz	0.8	3.0	3.1	4.1	370,621
Total Bolivian Lowlands	0.5	1.8	5.5	7.0	648,012
Total Land/Cattle[b]	3,272	11,737	3,557,772	4,551,901	
Ecuadorian Oriente	**1985**	**1995**	**1985**	**1995**	
Morona Santiago	0.2	0.9	4.8	6.9	24,606
Napo	1.3	1.6	2.3	2.2	37,682
Pastaza	0.2	0.4	0.8	0.9	29,137
Sucumbíos	0.1	2.5	2.8	2.6	16,014
Zamora Chinchipe	0.1	1.1	2.8	8.1	22,981
Total Ecuadorian Oriente	0.5	1.2	2.6	3.9	130,420
Total Land/Cattle	631	1,590	337,700	506,000	

Table 2.5. continued

State	Land area cultivated[a] (%) 1980s	1990s	Cattle/km² 1980s	1990s	Total land area[a]
Peruvian Selva		1994	1988	1997	
Amazonas	nd[c]	4.1	4.6	3.4	39,249
Loreto	nd	0.5	0.1	0.1	368,852
Madre de Dios	nd	1.0	0.2	0.4	85,183
San Martín	nd	7.6	1.8	2.2	51,253
Ucayali	nd	1.2	0.6	0.3	102,411
Total Peruvian Selva	nd	2.9	1.2	1.1	646,948
Total Land/Cattle	nd	18,529	759,000	688,265	

Sources: Brazil: 1985 data: IBGE (1990); 1996 data: IBGE (1998). Bolivia: 1988 crop data: INE (1995: 235–239); 1990 and 1999 agriculture data: INE (1999: 369–382); land area: INE (1997a: 5). Ecuador: 1985 data: MINAG (1986: 3, 38); 1995 data: INEC (1996: 7, 225); land area: INEC (2001). Perú: cropland data: INEI (1996b, vol. 1: 30, 39); cattle data: MINAG, cited in Webb and Baca (1999: 745); land area: INEI (1994: 48).

Notes

a: Cultivated land area and total land area are given in km². Cultivated land refers to land under annual and perennial crops and does not include pasture.

b: Total land area cultivated is given in km²; total cattle is given in head of cattle.

c: No data available.

expansion of land use helps account for the rise in deforestation. Interestingly, the proportion of land deforested as of 1996 is relatively high in the Oriente when compared to Brazil and Bolivia (compare Tables 2.1 and 2.2), though the percentage of cultivated land area and the cattle density are lower in the Oriente than the other two countries.

In the Peruvian selva, 1994 agricultural census data indicates cultivated land percentages that roughly track deforestation, as both are highest in Amazonas and San Martín. Estimates of cattle head in 1988 and 1997 show the same correspondence: cattle density is greatest in Amazonas and San Martín.

Contextual factors: National political economy

Across the Pan-Amazon basin deforestation reflects land-use profiles and changes. However, this explanation is not exhaustive. An alternative explanation for why population and net migration intermittently correspond with deforestation concerns the contrasting contexts of the countries that share the basin. It is therefore important to review additional relevant factors in order to provide an adequate basis for interpreting population–environment dynamics in the Pan-Amazon region.

In Brazil, the Legal Amazon has become an extremely complex and heterogeneous array of spaces containing numerous social groups and economic processes. In contrast to the 1970s, when Brazil's military regime had a pre-eminent influence upon frontier expansion, the crisis of the 1980s led to state retreat, involving the cessation of state-directed colonization and the withdrawal of many fiscal incentives (Binswanger 1991). Many interest groups, corporations, NGOs and other local and regional actors have asserted themselves, so that cattle, timber, and mineral interest groups increasingly contest land claims and resource use by indigenous groups, fishers, small farmers, landless peoples, and most famously, rubber tappers (Anderson 1990; Schmink and Wood 1992; Hall 1997). This has resulted in widespread deforestation in some areas where timber and cattle operations predominate (Wood and Skole 1998; Nepstad *et al.* 1999), and to a far lesser degree in areas where rubber tappers and other forest-dependent groups predominate (Hall 1997).

The countervailing social forces underlying deforestation that are emerging in the Brazilian Amazon necessarily respond to a changing national and global context. During the 1990s, the state adopted a "green" discourse when speaking about "development" in the Amazon, in part in response to international pressure and threats of frozen bank loans over deforestation in the region. In response, Brazil has created new environmental agencies, a new forestry code that requires 80 percent of private properties to be kept in forest, and a new National Integrated Policy for the Legal Amazon (Hall 1997). At the same time, the Avança Brasil program heralds the renewal of top-down infrastructure development projects in the Amazon, something reminiscent of policies from the 1970s that led to deforestation (Laurance *et al.* 2001). Avança Brasil, along with the reduction in inflation after the introduction of the Real in 1994, may be prompting new investment in the Amazon in the late 1990s. These changes may be generating a migration response, for net migration estimates presented here (Table 2.3) are less negative than estimates for 1991–1996 (Perz 2002a), implying renewed population gains due to migration during the late 1990s. This is occurring in an international context with the provision of bank loans for infrastructure and development projects in the Amazon, some driven by European demand for Brazilian soybeans (Hageman 1996). Whether due to national or international mechanisms, the recent deforestation estimates indicate a rise in annual forest clearing in the Brazilian Amazon from 1998 to 2000 (INPE 2002).

In Bolivia, much recent attention has focused on the neo-liberal reforms put in place since 1985. The so-called "first generation" of reforms in the late 1980s stabilized the national currency and secured preferential terms of trade with other Andean nations. This rendered key lowland exports such as soybeans and timber more profitable, and both products grew in importance manifold by the mid-1990s, largely due to foreign investment (Pacheco 1998: Chapter 6). The rise in deforestation due to these policies led to a "second generation" of reforms focused on the regulation of resource management, including new land-use and forestry laws in 1996, designed to both encourage more sustainable land-use practices and at the same time generate higher incomes (Pacheco 1998: Chapter 6; Llanque, Chapter 4). On balance, the effects of these two "generations" of reforms has been to increase deforestation, primarily by reinforcing a shift to mechanized soybean production in the region and encouraging unsustainable timber extraction, while not providing adequate enforcement of "second generation" regulations (Kaimowitz *et al.* 1999; Llanque, Chapter 4). It remains to be seen whether decentralization of the forestry law to municipal control will result in changes in land management practices (Kaimowitz *et al.* 1998). As in Brazil, most deforestation appears not to be caused by the activities of smallholders, who benefited little from the reforms. Smallholders primarily focus on labor-intensive and land-efficient coca production rather than land-extensive soybeans, cattle or timber (Pacheco 1998). As a result, population change plays a limited role in land-cover conversion in many areas. On the contrary, smallholders often sell their lands due to the "barbecho crisis", namely, declining productivity on repeatedly cleared plots (Thiele 1993). This allows soybean and cattle firms to expand, prompting rural–urban migration by poor families (Kaimowitz *et al.* 1999).

In Ecuador, an understanding of population–deforestation linkages can be facilitated by discussion of oil exploration-led frontier expansion, various social responses, and spillovers from Plan Colombia—the USA supported effort to eradicate coca production and trade. Land settlement in the Ecuadorian Oriente has proceeded slowly and over a long period of time, but accelerated in the 1970s following oil discoveries during the 1960s (Rudel and Horowitz 1993). Roads and drilling sites established by oil companies do account for some deforestation, but their effects on land cover are largely indirect, occurring via the resultant influx of poor populations to the Andes (Kimerling 2000). As elsewhere in the Amazon, land-use systems employed by smallholders have evolved over time, from annual plants to cash crops including coca and coffee, and now increasingly toward cattle farming (Pichón 1997). In addition to the corresponding migration that has led to further deforestation, indigenous groups in the Oriente have responded to environmental degradation in tribal territories (Uquillas 1989). While many groups have resisted oil exploration and its negative impacts in numerous capacities (Kimerling 2000), some have utilized the new roads to exploit forest resources, for example by ranching (Uquillas 1989). In contrast to the Amazon in Bolivia and parts of Brazil, land use in the Ecuadorian Oriente is not generally geared towards exports, in part due to political and economic instability (EIU 2000a). This has recently begun to change as Plan Colombia pushes drug production into northeastern Ecuador (Bannowsky 2001).

In Perú, regional and national changes are potentially modifying the place of the selva within larger society. Similar to other Andean countries, Perú's selva has experienced slow colonization by Andean families seeking additional land (Carpio 1988; Mora 1991; Limachi *et al.*, Chapter 3). This "Andeanization" of the selva often reflects seasonal migration strategies paired with Amazon land use designed to supplement Andean incomes (Collins 1988). However, this may also lead to permanent settlements and the expansion of deforestation over time (Dourojeanni 1990; Schjellerup 2000). During the 1980s and 1990s, the insurgent group Sendero Luminoso encouraged smallholder coca production in order to finance land settlement and support a leftist guerilla campaign against the Peruvian state. These activities fostered deforestation in remote parts of the selva (Bedoya and Klein 1996). The capture of insurgency leaders greatly reduced the influence of the Sendero Luminoso, but high and volatile coca prices have persisted, along with illicit deforestation. At the same time, cattle ranching emerged as an important activity in the selva (Varese 1999), largely in response to national economic growth during the 1990s and plans for a transoceanic highway through Brazil and Perú (CEPEI 2001). This occurred during the Fujimori presidency when Perú embarked on a strict neo-liberal economic program that encouraged oil exploration in the selva and growth in key export industries (EIU 2000b).

Conclusions

In order to address theoretical debates over population–environment linkages, this chapter presented a socioeconomic analysis of migration and land-cover change in the Amazonian states of Brazil, Bolivia, Ecuador and Perú. The general findings for the migration–deforestation relationship broadly support previous analyses that suggested limited or contingent effects. The analysis stepped beyond consideration of total migration, assessing rural and urban movement in addition to the labor-force composition of migrants and non-migrants. While this step afforded an explanation of the non-correspondence of total migration with land-cover change in some instances, it raised further questions about population–environment linkages.

The analysis therefore pursued two additional lines of inquiry, discussing land-use patterns as an intervening factor, and distinct national political economies as a contextualizing factor. The significance of land use as a mediating factor suggests that the migration–land cover link is indirect, and the disparities in national political economies indicate that the effect of migration on land cover is modified according to context. Thus, land use and political economy modify the effects of migration on land-cover change, thereby providing an explanation as to their limited correspondence. From this analysis it is evident that factors other than migration influence land-cover change.

The findings are further relevant to broader discussions about social processes and environmental change that have focused on the scale-specificity of the many factors involved. The insights afforded by consideration of migrant subgroups among rural and urban areas and labor-force categories, and the importance of land use and national political economies, both suggest that understanding of land-cover outcomes requires attention to assessment of aggregates of varying scales, whether state populations, population subgroups or national-level factors. This conclusion is consistent with documented impacts of land-cover change that operate on a range of levels, from local communities to the global economy (Perz 2002b; Wood 2002).

Notes

1. There are many ways to demarcate the Pan-Amazon basin, such as by hydrology, geology, vegetation, or language groups. Data limitations prevent consideration of the Guianas and Suriname. While the choice of countries and states here is somewhat artificial, excluding Colombia and Venezuela, the analysis does include a large majority of the land and people in the basin as it is defined by most criteria (Santana Nazoa 1991: Chapter 1).
2. Questions have been raised about the quality of the 1996 Population Count, which may have higher rates of undercount than the 1991 and 2000 Censuses, giving the false impression of a slowdown in population growth during the early 1990s, followed by a resurgence during the late 1990s (Caetano 2002).

References

Anderson, A.B., ed. 1990. *Alternatives to Deforestation*. New York: Columbia University Press.

Arizpe, L., M.P. Stone, and D.C. Major, eds. 1994. *Population and Environment: Rethinking the Debate*. Boulder: Westview Press.

Bannowsky, P. 2001. In the grip of the Anaconda: Plan Colombia, US drug war's effects in Ecuador. *NACLA Report on the Americas*, 35: 38–40.

Becker, B.K. 1995. Undoing the myths: The Amazon—an urbanized forest. In M. Clüsener-Godt and I. Sachs, eds., *Brazilian Perspectives on Sustainable Development in the Amazon Region*. Paris: UNESCO/ Parthenon.

Bedoya Garland, E., and L. Klein. 1996. Forty years of political ecology in the Peruvian upper forest: The case of the Upper Huallaga. In L.E. Sponsel, T.N. Headland, and R.C. Bailey, eds., *Tropical Deforestation: The Human Dimension*, 165–186. New York: Columbia University Press.

Bilsborrow, R.E. 1987. Population pressures and agricultural development in developing countries: A conceptual framework and recent evidence. *World Development*, 15(2): 183–203.

Bilsborrow, R.E. 2002. Migration, population change, and the rural environment. *Environmental Change and Security Project Report*, 8: 69–94.

Binswanger, H.P. 1991. Brazilian policies that encourage deforestation in the Amazon. *World Development*, 19: 821–829.

Binswanger, H.P. and V. Ruttan. 1978. *Induced Innovation, Technology, Institutions and Development*. Baltimore: Johns Hopkins University Press.

Boserup, E. 1965. *The Conditions of Agricultural Growth*. London: Allen and Unwin.

Browder, J.O. and B.J. Godfrey. 1997. *Rainforest Cities: Urbanization, Development and Globalization of the Brazilian Amazon*. New York: Columbia University Press.

Caetano, A.J. 2002. Personal communication. Belo Horizonte, Brasil: Universidade Federal de Minas Gerais.

Carpio, J. 1988. Settlement policies in the forest highlands of Perú. In A.S. Oberai, ed., *Land Settlement Policies and Population Redistribution in Developing Countries*, 355–386. New York: Praeger.

CEPEI (Centro Perúano de Estudios Internacionales). 2001. *La Integración Regional entre Bolivia, Brasil y Perú*. Lima: CEPEI.

Collins, J. 1988. *Unseasonal Migrations: The Effects of Rural Labor Scarcity in Perú*. Princeton: Princeton University Press.

Davis, K. 1963. The theory of change and response in modern demographic history. *Population Index*, 29(4): 345–366.

Dourojeanni, M. 1990. *Amazonía: ¿Que Hacer?*. Iquitos: CETA.

EIU (Economist Intelligence Unit). 2000a. *Country Profile, Ecuador. 1999–2000*. London: EIU.

EIU. 2000b. *Country Profile: Perú. 1999–2000*. London: EIU.

Fearnside, P.M. 1990. Environmental destruction in the Brazilian Amazon. In D. Goodman and A. Hall, eds., *The Future of Amazonia*, 171–220. London: St. Martin's Press.

Gash, J.H.C., C. Nobre, J.M. Roberts, and R.L. Victoria. 1996. *Amazonian Deforestation and Climate Change*. Chichester: John Wiley and Sons.

Geist, H.J. and E.F. Lambin. 2002. Proximate Causes and Underlying Forces of Tropical Deforestation. *BioScience*, 52(2): 143–150.

Gibson, C., E. Ostrom, and T-K. Ahn. 2000. The concept of scale and the human dimensions of global change: A survey. *Ecological Economics*, 32: 217–239.

Goldscheider, C. 1983. Modernization, migration, and urbanization. In P.A. Morrison, ed., *Population Movements*, 47–66. Liege: Ordina/IUSSP.

Hageman, H. 1996. *Bancos, Incendiários, e Florestas Tropicais*. Rio de Janeiro: FASE/ IBASE/ ISA.

Hall, A.L. 1992. Putting people last: A sociological perspective on development policies for Brazilian Amazonia and the question of sustainability. In W. Pansters, ed., *Amazonia: Ecology and sustainable development*, 110–126. Utrecht: ISOR/ University of Utrecht.

Hall, A.L. 1997. *Sustaining Amazonia: Grassroots Action for Productive Conservation*. Manchester: Manchester University Press.

Hogan, D.J. 1991. Crescimento demográfico e meio ambiente. *Revista Brasileira de Estudos Populacionais*, 8(1–2): 61–71.

IBGE (Instituto Brasileiro de Geografia e Estatística). 1990. *Censo Agropecuário. 1985*. Rio de Janeiro: IBGE.

IBGE. 1991. *Sinopse Preliminar do Censo Demográfico. 1991*. Rio de Janeiro: IBGE.

IBGE. 1996. *Censo Demográfico 1991: Microdados da Amostra*. Rio de Janeiro: IBGE.

IBGE. 1998. *Sistema IBGE de Recuperação Automática (SIDRA)—Censo Agropecuário—1995/96*. www.sidra.ibge.gov.br.

INE (Instituto Nacional de Estadística). 1992. *Censo Nacional de Población y Vivienda de 1992: Resultados Finales*. La Paz: INE.

INE. 1995. *Anuário Estadístico 1995*. La Paz: INE.

INE. 1997. *Características Demográficas de la Población en Bolivia*. La Paz, INE/ UNFPA/ CELADE.

INE. 1999. *Anuário Estadístico 1999*. La Paz: INE.

INEC (Instituto Nacional de Estadística e Censos). 1992. *V censo de Población y IV de Vivienda. 1990*. Quito: INEC.

INEC. 1996. *Encuesta Nacional de Superficie y Producción Agropecuário de 1995*. Quito: INEC.

INEC. 2001. *Proyección de la Población Total por Años Calendario según Provincias*. http://www.inec.gov.ec/censo/proyecc/proye01.htm.

INEI (Instituto Nacional de Estadística e Informática). 1994. *Censos Nacionales 1993 IX de Población e IV de Vivienda: Resultados Definitivos*. Lima: INEI.

INEI. 1996. *Censo Nacional de Agropecuário: Resultados Definitivos*. Lima: INEI.

INEI. 1997. *Perú: Estadísticas del Medio Ambiente*. Lima: INEI.

INPE (Instituto Nacional de Pesquisas Espaciais). 2002. Monitoramento da Floresta Amazônica por Satélite, Relatório 2000–2001. www.inpe.br/Informacoes_Eventos/.

Kaimowitz, D. and A. Angelsen. 1998. *Economic Models of Tropical Deforestation: A Review*. Jakarta: CIFOR.

Kaimowitz, D., C. Vallejos, P. Pacheco, and R. Lopez. 1998. Municipal governments and forest management in Lowland Bolivia. *Journal of Environment and Development*, 7(1): 45–59.

Kaimowitz, D., G. Thiele, and P. Pacheco. 1999. The effects of structural adjustment on deforestation and forest degradation in Lowland Bolivia. *World Development*, 27(3): 505–520.

Kimerling, J. 2000. Oil development in Ecuador and Perú: Law, politics and the environment. In A. Hall, ed., *Amazonia at the Crossroads: The Challenge of Sustainable Development*, 73–96. London: Institute for Latin American Studies.

Kosinski, L. 1992. *Ecological Disorder in Amazonia: Social Aspects*. Rio de Janeiro: ISSC.

Lambin, E.F. *et al.* 2001. The causes of land use and land-cover change: Moving beyond the myths. *Global Environmental Change*, 11: 261–269.

Laurance, W.F. *et al.* 2001. The future of the Brazilian Amazon. *Science*, 291: 438–439.

Liverman, D., E.F. Moran, R.R. Rindfuss, and P.C. Stern, eds. 1998. *People and Pixels: Linking Remote Sensing and Social Science*. Washington, DC: National Academy Press.

Malthus, T.R. 1989(1798). *An Essay on the Principle of Population.* Cambridge: Cambridge University Press.

Martine, G. 1994. A redistribuição espacial da população Brasileira durante a década do 80. *IPEA Texto para Discussão No. 329*, Rio de Janeiro.

Massey, D.S., J. Arango, G. Hugo, A. Kouaouci, A. Pellegrino, and J.E. Taylor. 1993. Theories of international migration: A review and appraisal. *Population and Development Review*, 19(3): 431–466.

Mazur, L.A. 1994. *Beyond the Numbers.* Washington, DC: Island Press.

MINAG (Ministerio de Agricultura). 1986. *Estado de la Superficie Cosechada e de la Producción Agrícola del Ecuador.* Quito: MINAG.

Mora, C. 1991. Expansión de la frontera agrícola en la Amazonía Peruana. In L. Munévar, ed., *Colonización del Bosque Húmedo Tropical*, 125–146. Santafé de Bogotá, DC, FPC/ COA.

Nepstad, D. *et al.* 1999. Large-scale impoverishment of Amazonian Forests by logging and fire. *Nature*, 398: 505–508.

Ness, G.D., W.D. Drake, and S.R. Brechin, eds. 1993. *Population–environment Dynamics.* Ann Arbor: University of Michigan.

Ozório de Almeida, A.L. and J.S. Campari. 1995. *Sustainable Settlement in the Brazilian Amazon.* Oxford: Oxford University Press.

Pacheco, P. 1998. *Estilos de Desarrollo, Deforestación de los Bosques en las Tierras Bajas de Bolivia.* La Paz: CIFOR/ CEDLA/ TIERRA.

Panayotou, T. 1996. An inquiry into population, resources, and environment. In D.A. Ahlburg, A.C. Kelley, and K. Oppenheim Mason, eds., *The Impact of Population Growth on Well-being in Developing Countries*, 259–298. New York, Springer-Verlag.

Pebley, A.R. 1998. Demography and the environment. *Demography*, 35: 377–389.

Perz, S.G. 2002a. Population growth and net migration in the Brazilian Legal Amazon, 1970–1996. In C.H. Wood and R. Porro, eds., *Deforestation and Land Use in the Amazon*, 107–129. Gainesville: University of Florida Press.

Perz, S.G. 2002b. The changing social contexts of deforestation in the Brazilian Amazon. *Social Science Quarterly*, 83(1): 35–52.

Perz, S.G., C. Aramburú, and J. Bremner. 2002. Population and deforestation in the Pan-Amazon Basin: A comparison of Brazil, Bolivia, Colombia, Ecuador, Perú and Venezuela. Manuscript under journal review.

Pfaff, A.S.P. 1999. What drives deforestation in the Brazilian Amazon? *Journal of Environmental Economics and Management*, 37: 26–43.

Pichón, F.J. 1996. The forest conversion process: A discussion of the sustainability of predominant land uses associated with frontier expansion in the Amazon. *Agriculture and Human Values*, 13: 32–52.

Pichón, F.J. 1997. Settler households and land-use strategies in the Amazon Frontier: Farm-level evidence from Ecuador. *World Development*, 25(1): 67–91.

Ritchey, P.N. 1976. Explanations of migration. *Annual Review of Sociology*, 2: 363–404.

Rodriguez, F. 2001. Amazonia: Una herencia al futuro. Paper presented at the international seminar entitled *Cambios Demográficos, Económicos y Sociales y su Impacto Ambiental en la Amazonía Andina*. Iquitos: June 11–13, 2001.

Rudel, T.K. and B. Horowitz. 1993. *Tropical Deforestation: Small Farmers and Land Clearing in the Ecuadorian Amazon*. New York: Columbia University Press.

Santana Nazoa, A. 1991. *La Cuestión Científica y Tecnológica en el Amazonas Venezolano: Evaluación y Perspectivas*. Caracas: CENDES.

Schjellerup, I. 2000. La Morada: A case study on the impact of human pressures on the environment in the Ceja de Selva, northeastern Perú. *Ambio*, 29(7): 451–454.

Schmink, M. and C.H. Wood. 1992. *Contested Frontiers in Amazonia*. New York: Columbia University Press.

Serrão, E.A.S. and A.K.O. Homma. 1993. Brazil. In *Sustainable Agriculture and the Environment in the Humid Tropics*, 265–331. Washington, DC: National Academy Press.

Shryock, H.S. and J. Seigel, *et al.* 1976. *The Methods and Materials of Demography, Condensed Edition*. Washington, DC: US Government Printing Office.

Thiele, G. 1993. The dynamics of farm development in the Amazon: The *Barbecho* crisis model. *Agricultural Systems*, 42: 179–197.

Torres, H. and H. Costa. 2000. *População e Meio Ambiente: Debates e Desafios*. São Paulo: Editora SENAC.

Turner, B.L. II, G. Hyden, and R.W. Kates, eds. 1993. *Population Growth and Agricultural Change in Africa*. Gainesville: University of Florida Press.

Turner, B.L. II, W.C. Clark, R.W. Kates, J.F. Richards, J.T. Mathews, and W.B. Meyer, eds. 1990. *The Earth as Transformed by Human Action*. Cambridge: Cambridge and Clark Universities.

UN (United Nations). 1970. *Manuals on Methods of Estimating Population, Manual VI: Methods of Measuring Internal Migration*. New York: UN.

Uquillas, J. 1989. Social impacts of modernization and public policy, and prospects for indigenous development in Ecuador's Amazonia. In D.A. Schumann and W.L. Partridge, eds., *The Human Ecology of Tropical Land Settlement in Latin America*, 407–431. Boulder, Westview Press.

Varese, M. 1999. *Drivers of Investment in Cattle among Landholders in the Southern Peruvian Amazon*. Master's thesis, University of Florida.

Walker, R.T., S.G. Perz, M. Caldas, and L.G. Teixeira da Silva. 2002. Land use and land-cover change in forest frontiers: The role of household life cycles. *International Regional Science Review*, 25(2): 169–199.

Wood, C.H. 1982. Equilibrium and historical-structural perspectives on migration. *International Migration Review*, 16(2): 298–319.

Wood, C.H. 1992. Demographic perspectives on ecological disorder in Brazilian Amazonia. In L. Kosinski, ed., *Ecological Disorder in Amazonia: Social Aspects*, 129–148. Rio de Janeiro, ISSC.

Wood, C.H. 2002. Land use and deforestation: Introduction. In C.H. Wood and R. Porro, eds., *Deforestation and Land Use in the Amazon*, 1–40. Gainesville: University of Florida Press.

Wood, C.H. and D.L. Skole. 1998. Linking satellite, census and survey data to study deforestation in the Brazilian Amazon. In D. Liverman, E.F. Moran, R.R. Rindfuss, and P.C. Stern, eds., *People and Pixels: Linking Remote Sensing and Social Science*, 70–93. Washington, DC: National Academy Press.

3
Models of Migration in the Peruvian Amazon and their Impact on Tropical Forests

Luis Limachi, Wil de Jong and Carlos Cornejo

Introduction

Discussions during the 1980s and 1990s on the deforestation of the tropical Amazon proposed a simple formula that related population density increases at agricultural frontiers to progressive deforestation. More recently, however, researchers and academics have begun to seriously question this simplistic correlation. Some argue that the type of resource use practiced at the forest frontier and the availability of technology—and by extension availability of capital investment—are important factors in determining how growing populations affect agriculture–forest frontiers. The subject of migration is closely related to this discussion for several reasons. First, migration influences population density and the subsequent demand for various types of natural resources, including forestland. Second, population characteristics may influence resource use practices and/or access to technology and capital (Bradley, Chapter 5).

Migration has played an essential role in the settlement of the Amazonian lowlands of Perú. The region's first inhabitants migrated there over ten thousand years ago. Settlement has been a key factor creating connections between the Amazon region and both modern Peruvian society and the rest of the world. In this chapter we particularly analyze how the incorporation of Amazonian Perú into mainstream Peruvian society has developed throughout much of the twentieth century, focusing particularly upon the last few decades. We also examine the links between migration and changing forest cover in the Peruvian Amazon.

In this chapter we not only describe the links between migration and deforestation, but also illustrate diverse factors that influence processes of migration. The forces under consideration here are either pull-factors (government policies, for example road construction, or economic opportunities at the migrants' destination) or push-factors (conditions at the migrants' point of origin, production and trade of illicit crops, and guerilla group activity).

The Bulk of the chapter is devoted to analysis of the processes of migration in relation to the Peruvian Amazon, taking into account the factors that influence these processes. Three specific cases of migration and how these relate to deforestation will be discussed. These cases include: migration to the department of San Martín caused by the construction of the Carretera Marginal (literally "marginal highway"), a road constructed in the foothills of the Andes; migration related to the Federico Basadre road that connected the city of Pucallpa with Lima; and migration influenced by the construction of the Iquitos to Nauta road in the heart of the Peruvian lowland Amazon. Data was collected during much of the 1990s with some of the findings published elsewhere (Limachi 1995, 2002). In addition to this body of work, we draw heavily upon other studies conducted on each of these three cases.

The chapter is divided into five sections. In section two we summarize several contemporary issues concerning migration and its impact upon the wellbeing of populations and the environment. In section three we discuss general aspects of migrations in the Amazon region of Perú linking these waves to economic booms, state policies, and conditions in the migrants' places of origin. In section four we present the three specific case studies mentioned above. In the final section we discuss some of the implications of our analysis.

Theories on migration

The discourse concerning migration and its effect on tropical forests can be divided into two central themes; the causes and impacts of migration. The former addresses questions such as: What are the forces that encourage groups of people to transfer part or all of their lives from one place to another? The latter focuses on the impact that migration has on the environment, the immigrants themselves, those who reside in the migrants' destination, and other related elements.

Resource shortage is generally considered to be the primary underlying cause of population movement (Bilsborrow 2002). People who migrate towards tropical forest zones experience some scarcity of resources at home. This may be land or capital scarcity. In order to fully comprehend the motivations for migration, however, it is important to understand the opportunities perceived by the migrants before initiating their journey. Migration will only occur if the migrants recognize differences between their home and destination. These disparities may be in the form of economic opportunities, or other benefits such as political security. Mechanisms of information exchange, the type of information exchanged, and access to this information by people considering migrating are all important factors determining migration dynamics. Migrants' perceptions are an important driving force. The capacity of potential migrants to respond to perceived opportunities is an additional factor influencing migrant behavior (Bilsborrow 2002).

At an agricultural frontier, the impact of immigrants on their place of destination is similar to the effects resulting from increased population density. However, immigrants very often bring with them special attributes that are equally important in explaining changes that take place once they arrive at their destination. They may offer new skills once they arrive, or they may lack necessary skills for survival. Their cultural attitudes towards collaboration, forming networks and other social interactions, may differ from those of other inhabitants. These factors greatly influence the types of resource use they practice, how their choices affect their quality of life and that of others, and the overall impacts upon the environment.

At this point, we have only a limited understanding of the key factors that influence both migration and the impact that migration has on places of destination. Recent research suggests that increased population density arising from migration is related to deforestation through a number of intermediate factors such as land use and local economic and political structures (Perz *et al.* 2002; Perz, Chapter 2). For example, in Latin America, expansion of pasture is the principal cause of deforestation which in turn is as closely related to migration as to other factors that favor using land for cattle production (Bilsborrow 2002). In order to explain why migration has caused environmental degradation, some have emphasized the crucial role of factors such as natural resources endowment, the presence and effectiveness of communal or national institutions, regional, national, and international policies, markets, and cultural factors (Bilsborrow 2002). Household, community, and national level factors both influence migration and play an important role in the impact of migration on tropical forest areas, although the exact nature and importance of these factors remains unclear (Bilsborrow 2002). The scale of observation used to analyze migration and its impact is a crucial variable that can influence results (Perz *et al.* 2002).

Migrations and the Peruvian Amazon

The first waves of migration to the Amazon in modern history began during the colonial period as a result of the first military and religious incursions of the Spaniards. The flow of migrants in this period was limited. At that time the population was located in and around some eighty centers, the most important of which were Moyobamba founded in 1539, Chachapoyas in 1566, Borja in 1634, Jeberos in 1640, and Yurimaguas in 1709. Estimates for the population of the province of Maynas, which at that time included the largest part of the Peruvian Amazon, hovered around an average of approximately 25 thousand inhabitants. This estimate excludes a large proportion of the native indigenous population that was dispersed throughout the Amazon region and therefore not registered. Intensive waves of migration took place during the several booms and busts of commercially exploited forest products. The most well-known is the exploitation of rubber that occurred

Figure 3.1. Map of the locations referred to in the chapter

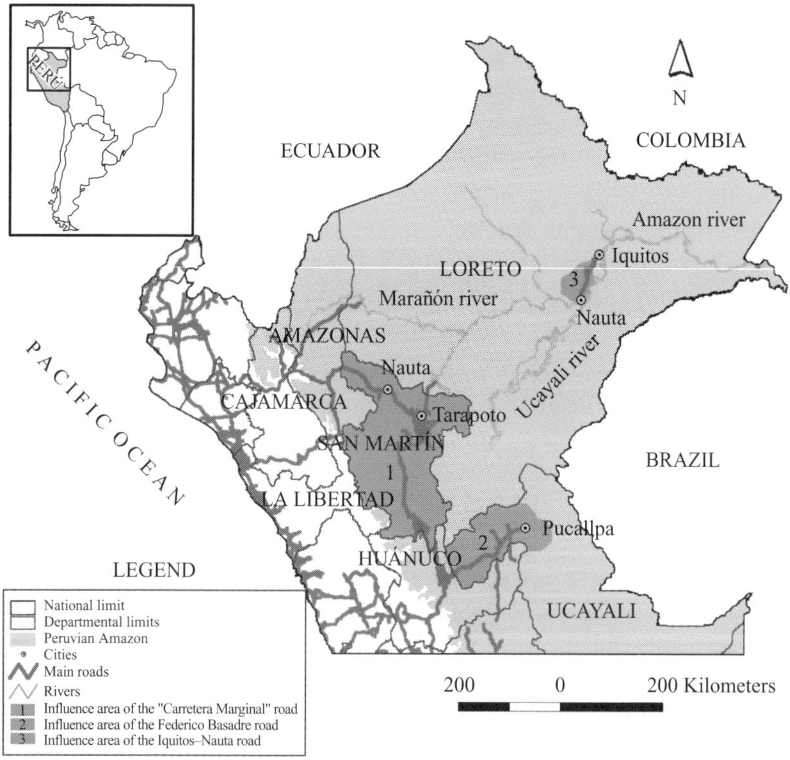

roughly between 1890 and 1920. Before and after that period, other forest products such as quinine (*Cinchona officinalis*), barbasco (*Lonchocarpus nicou*)—a natural pesticide—and later oil were exploited. In the last two decades of the twentieth century, three divergent trends of migration can be identified: 1) rural–urban migration led to the growth of the main cities in the Amazon region during much of the last four decades of the twentieth century; 2) emigration of the Amazonian population towards the coast, primarily in the direction of Lima, has continued during the last decades; and 3) recent influences on migration have been the activities of guerilla groups such as Sendero Luminoso, the extreme poverty in Andean communities, and the drug trade in the Andean foothills.

According to available statistics, in 1993, more than 320 thousand inhabitants from the departments of Amazonas, Madre de Dios, Loreto, San Martín, and Ucayali had been born outside the Amazonian region (that is, had migrated to the region at some point during their lives. See Figure 3.1 for locations of places mentioned in this chapter). Since 1981 the population of migrants had grown by approximately 170 thousand people, or just over 50 percent of the registered immigrant population in 1993. This increase coincides with the rise of extreme poverty in the Andes region and also with activities of guerilla groups and drug trafficking.

During the 1981–1993 inter-census period the cities of the Amazon experienced an unusual rate of growth. The city with the largest population growth during this period was Puerto Maldonado, with a population increase of about 7.8 percent annually. Table 3.1 provides similar data for other major cities. During this time, the Amazonian population had an average annual

Table 3.1. Most populated cities in the Peruvian Amazon

City	Population 1981	Population 1993	Inter-census increase: Total increase	Inter-census increase: %	Yearly average growth rate
Puerto Maldonado	12,693	31,249	18,556	146.2	7.8
Tarapoto	34,979	7,783	2,804	22.4	6.9
Pucallpa	89,604	172,286	82,682	92.3	5.6
Huánuco	61,812	18,814	57,002	92.2	5.6
Moyobamba	14,376	24,800	10,424	72.5	4.6
Iquitos	178,738	274,759	96,021	53.7	3.6
Chachapoyas	11,853	5,785	3,932	33.2	2.4
Total	404,055	715,476	311,421	77.1	4.9
Lima Metropolitana	4,573,227	6,321,173	1,747,946	38.2	2.7

Source: Limachi (1995).

Table 3.2. Internal migration flows in the Peruvian Amazon, 1981–1993

Department of origin	Department of destination					
	Amazonas	Loreto	Madre de Dios	San Martín	Ucayali	Total
Amazonas		1,779	21	22,661	626	25,087
Loreto	543		271	12,883	39,342	53,039
Madre de Dios	12	308		92	725	1,137
San Martín	2,046	21,855	156		13,970	38,027
Ucayali	75	9,039	257	2,900		12,271
Total	2,676	32,981	705	38,536	54,663	129,561

Source: Limachi (1995).

increase of 4.9 percent. This figure is larger than the 2.7 percent annual growth rate of metropolitan Lima for the same period. The high rates of urban growth in the Amazon are explained by rural–urban migration, although extra-regional migration and natural growth also contributed. The decline of the rural population relative to the urban one corroborates this conclusion. In 1973 the rural population was 60 percent of the total regional population; this number diminished to 50 percent in 1993.

Interdepartmental migrations within Amazonian Perú are primarily oriented in three directions: 1) from Amazonas to San Martín; 2) from San Martín to Loreto and Ucayali; and 3) from Loreto to Ucayali. As can be observed in Table 3.2, the department of Ucayali is the most popular destination for regional migration, while the department that loses the greatest number through migration is Loreto. The factors that determine the direction of migratory flows within the Amazon territory remain insufficiently understood. Nevertheless, in the specific case of migration to Ucayali, this is undoubtedly related to the commercial and industrial development of the city of Pucallpa during the last thirty years as a result of the construction of a road connecting it with the coast.

The booms of natural products and the processes of migration

Since Peruvian independence in 1821, migration to the Peruvian Amazon has been linked to changes in the regional economy caused by the trade of products such as rubber, barbasco, wood, and oil. The rubber boom began when the first exports were made from the country in 1862, continuing until approximately 1920. Large numbers of migrants were attracted from the north of the country and investors came from abroad to exploit rubber. The recent arrivals occupied numerous small rubber outposts. The population of the Amazon, which had remained fairly constant prior to the rubber

Table 3.3. Evolution of the population of the departments Amazonas and Loreto 1862–1896

Department	1862	1876	1896	Increase 1862–1896 (%)
Amazonas	32,595	34,284	70,676	117
Loreto	52,124	61,125	100,596	93

Source: Rodriguez (1986).

boom, began to increase tremendously during that period. Between 1862 and 1896, the two departments of Amazonas and Loreto, the most important departments of the Peruvian Amazon, increased their population by 117 percent and 93 percent respectively (Table 3.3).

After the decline of the rubber boom beginning in 1910, the Amazon continued to receive migrants due to the improvement in communications between the coast and the Amazon. Later, after the discovery of oil in the region, thousands of workers were required for the exploration teams. Finally, during the 1980s the illegal production of coca leaves caused a subsequent economic boom particularly in the main cities such as Iquitos, Pucallpa, and Tarapoto (San Roman 1994). The production of coca leaves and of cocaine had begun in the 1970s but did not become widespread in the foothills of the Andes until the 1980s. Informal coca plant producers invaded the entire river basins of the Huallaga, Aguaytía, Apurimac, and Pachitea. This boom paralleled the emergence of guerilla groups. The latter did not displace illicit production or depopulate the region. On the contrary, rebels and narcotics traffickers established an alliance and both thrived during the 1980s.

Policies and migration

Policies promoting migration into the Peruvian Amazon have existed since the inception of the republic. The Peruvian government promulgated a series of laws and decrees, beginning with a law on 21 November, 1823, that gave free ownership titles to all nationals and foreigners who settled in the Amazon (Laburre and Correa, cited in San Roman 1994). After the rubber boom, policies focused on the construction of road networks in order to encourage settlement in the region. In 1933 the construction of the Lima–Pucallpa highway was approved. This project began in 1937 and was completed in 1943. The construction of roads, and later additional policy incentives to promote investment in the Amazon, stimulated large scale migration, particularly from the departments of Huánuco and Junín to the upper and medium reaches of the Huallaga River. In successive national censuses from 1940, the rates of population growth in that region were

Table 3.4. Population indicators of the Peruvian Amazon according to national census[a]

Year	Number of inhabitants	% of national population	% of urban population	Amazon inter-census growth	Perú inter-census growth
1940	414,452	6.7			
1961	865,210	8.7	3.6	1.9	
1972	1,341,922	9.9	39.3	4.1	2.8
1981	1,813,887	10.7	42.2	3.4	2.6
1993	2,800,100	12.4	49.3	3.8	2.0

Source: INEI (2002).

a: Calculations based on the population change of the departments of Amazonas, Loreto, Madre de Dios, San Martín and Ucayali.

always superior to national growth rates (Table 3.4). The contribution of the Amazonian population to the national population increased from 6 percent in 1940 to 12 percent in 1993. In the first years of the 1970s the Carretera Marginal was completed. This road crosses the region from north to south. Since then, additional feeder roads have been constructed.

The first post-World War II policy, the "Economic and Industrial Protection Law", intended to promote economic activities in the Peruvian lowlands. In 1965, Law No. 15600 declared the Amazon region a tax free zone for fifteen years. In 1970, the Law of Industries (no. 18350) granted a series of financial incentives for industries in the Amazon region. The Organic Law of Industries (no. 23407) exempted Amazon-based industries from paying several kinds of taxes. In 1998 the Investment Promotion Law (no. 27037) also exempted payment of taxes under certain conditions. Although these laws may not have strictly intended to promote migration into the region, they did have that effect since investment improves labor opportunities which in turn motivate migration.

The government undertook more direct interventions in the form of colonization projects designed to complement agrarian reforms that took place in the 1970s. In 1979 the government initiated the Special Project Central and Lower Huallaga that promoted agricultural development and frontier expansion. Two years later the Special Project Upper Huallaga was created, extending the objectives of the previous project to grant tenure security to landholders (Salazar 1984). Between 1983 and 1984, four other special projects were created with the financial support of the US Agency for International Development (USAID) and the World Bank. These initiatives were identified with the name of the location where they took place: Alto Mayo, Pichis-Palcazu, Satipo-Chanchamayo, and Jaen-San Ignacio. These efforts are evidence of the Peruvian state's clear intention to mobilize populations from underdeveloped parts of the country to move into these regions.

In addition to the above mentioned activities, the government established a program during the 1980s to provide credit to farmers at zero nominal rates of interest. This project contributed substantially to the occupation of land and deforestation, as it led to the large-scale establishment of government-subsidized production of bovine livestock, rice, and corn.

Unquestionably, another determining factor for massive occupation of the lowland region was, and continues to be, agricultural prices in national and international markets. Until the mid-1960s, the region relied heavily upon the production of coffee, tobacco, and cotton. When the international prices of these products declined, economic stagnation and a reverse flow of migration took place. Only after finalizing the Carretera Marginal, and as a result of the subsidies implemented by the State, did the boom of rice and corn production between 1977 and 1989 occur. This was additionally triggered by the wide adoption of irrigation technology and the support provided by the projects discussed above. Each of these factors helped drive regional agricultural prosperity during that period. At the end of the 1980s, when agricultural credits were suspended and guerilla groups spread throughout the region, corn and rice production declined steeply.

The origin of immigrants

Through the examination of census data it is possible to determine the changes that have taken place in the centers of migrant origins. These changes appear to coincide with variations in socioeconomic conditions and communication capabilities of these centers. Until 1970, migrants moving to departments in the Amazon region came primarily from other departments in the same region and from the coast—particularly from the department of Loreto, which during preceding periods had maintained close commercial ties with adjacent regions. This was facilitated by the fluvial connections it had with the rest of the Peruvian Amazon. During the early twentieth century Loreto went through a generalized economic crisis, a product of the collapse of the rubber boom. As a result, it became an exporter of people who largely migrated towards the department of San Martín. The immigrants from the coast originated primarily from the departments of Lambayeque and La Libertad. This migration was facilitated largely by the preexisting commercial links between these regions and by the beginning of air transportation from Chiclayo and Trujillo (capital of the departments of Lambayeque and La Libertad respectively) to several points in the Amazon region.

Table 3.5. Immigrant population of the Peruvian Amazon according to department of origin, 1993

Department of Origin	Immigrants (1993) Inhabitants	%
Cajamarca	121,053	37.8
Lima and Callao	37,292	11.7
Huánuco	34,855	10.9
Piura	22,174	6.9
Cusco	19,770	6.2
La Libertad	18,882	5.9
Lambayeque	17,042	5.3
Junín	12,682	4.0
Ancash	8,770	2.7
Other departments	27,552	8.6
Total immigrants	320,072	100.0

Source: Limachi (1995).

From the 1970s onward, the Andes departments began to contribute substantially to the migration flows towards the Amazon region. For example, in 1940 only 11 percent of the immigrants to San Martín came from the Andes region. In the following decades the proportions of immigrants from the Andes into San Martín were 35 percent, 47 percent, and 51 percent respectively. These flows are clearly correlated with the construction of the Carretera Marginal and more generally with the structural changes that occurred in Perú during that time. The flow of immigrants from the south decreased and was replaced by a flow from the north. This was closely linked to the advance of the Carretera Marginal into that region and its connection to the Pan-American Highway. Upon completion of these connections, migration from the department of Cajamarca increased tremendously, representing close to 35 percent of the total immigrants into the department of San Martín.

Recently, migration into the Amazon region has been largely occurring from neighboring departments, particularly those connected by road (Table 3.5). More than 50 percent of immigrants come from the Andes regions; 26 percent are from the coast, and 22 percent from the departments of the Amazon region. The migratory flows originating from the Andes are predominantly of the rural–rural type (see INEI 1995; Aramburú 1982; Martínez 1969). Migration into the Amazon region functioned as a mechanism to resolve land shortage problems in the Andes region. However, immigrants originating from the coast and Amazon departments consisted mainly of traders, state officials, and urban laborers.

Figure 3.2. Age composition of the migrating population to the Amazon region, 1993

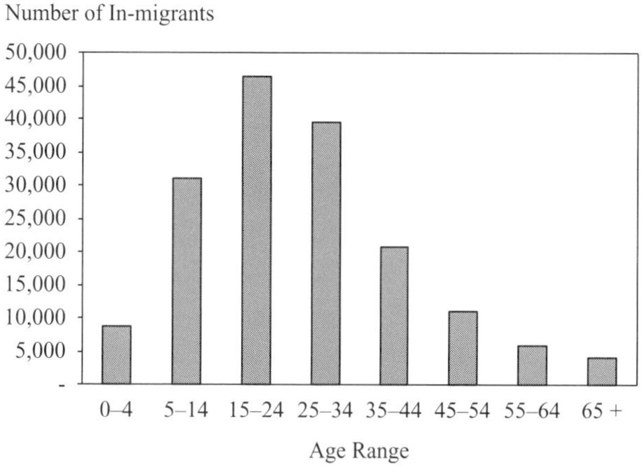

A greater number of males than females participate in the migratory flows to the Amazon (Aramburú 1982; Martínez 1969; INEI 1993). This is reflected in the composition of the population of immigration areas where males persistently form the majority. Younger age groups are also relatively well represented. Although the censuses do not record the exact age of migrants, the age-structure of migrants at the time of census suggests that those who migrate are between the ages of fifteen and thirty (Figure 3.2). These same characteristics were reported by the studies of Martínez (1969) and Aramburú (1982).

Conditions in the centers of origin

The socioeconomic processes within the Andean region that contributed to migration in the last half of the twentieth century include population explosion, agrarian reform, increasing levels of poverty, and the emergence of guerilla groups. The population explosion had its origin in the sudden nationwide decline in mortality rates. As a result, annual population growth rates increased beyond 2 percent per year. This affected the Andean region particularly, due to its coincidence with the absence of important sources of work outside the farming sector, the concentration of the country's incipient industries on the coast, and the continued existence of feudal forms of land ownership. The growth of the Andean population initially resulted in an increased pressure on resources and eventually led to various waves of emigration towards the coastal region, particularly Lima and the Amazon region (Martínez 1969).

Another factor that decisively influenced migration towards the Amazon region was the implementation of agrarian reform in 1969. Agrarian reform affected large properties on the coast and in Andean regions in particular. This process ended the feudal system of land ownership and liberated an important labor mass formerly tied to the land. More recently, the processes of impoverishment and the emergence of guerilla groups in the Andean region have contributed to increased migratory flows from the Andes. The rural Andes and the Amazon region both exhibit poverty rates of over 35 percent as opposed to 26 percent in coastal regions (Verdera 2000). Simultaneously, during the 1980s activities of guerilla groups centered primarily on these same Andean regions. These activities led to the displacement of 600 thousand farmers from the Andes towards the cities of the coast and the Amazon.

Migration and deforestation: Three cases from Perú

The "Carretera Marginal" and migration to San Martín

The San Martín region is located in the north-eastern part of Perú in the foothills of the Andes. The region has experienced several waves of migration during the last century, reflected in the large-scale occupation and expansion of the agricultural frontier and related deforestation. The impact of these waves is mirrored in the population growth which multiplied by 1.7 times between 1940 and 1960 and 5.8 times between 1960 and 1993. The total increase in the population of the Amazon during this period was 4.2 times. In just twenty-three years some 960 thousand ha of primary forest was converted to agricultural or other non-forest land uses. The occupation of San Martín in the last thirty years is closely related to the development of road infrastructure, chiefly the Carretera Marginal, and influenced by the special projects mentioned previously.

Construction of the Carretera Marginal began in the middle of the 1960s. In 1968 it connected Tarapoto with the Upper Mayo, and the following year with Juanjui. In the middle of the 1970s, the Upper Mayo became integrated with the north of the country, and by the end of the decade it was connected to Tingo Maria and Juanjui. The Carretera Marginal changed the commercial routes from San Martín towards the markets of the coast (ITDG 1993). These changes had a determining impact upon the demographic of flows towards San Martín.

In 1940, a total of 7.7 percent of the population of San Martín had migrated to coastal markets. This portion increased to 13.2 percent in 1961, when the total population had increased from 94 thousand to 161 thousand inhabitants.

During that time, and for previous periods, the commercial cultivation of coffee and cotton had become of primary importance to the region. Throughout the inter-census period of 1961–1972, however, commercial agriculture underwent a crisis due to the drop in international market prices. The result was the stagnation of agricultural production and the emigration of the population towards other regions.

After the 1970s, migration to the region reversed again, coinciding with the opening of the Carretera Marginal. The 1981 census report showed 75,096 recent immigrants and the 1993 census report 175,363, representing 31.7 percent of the total population of San Martín. Although reliable information is available only to the year 1993, population estimates conducted by INEI (2002) suggest that the trend to migrate to San Martín seems to have stabilized. Nevertheless, within the region a process of consolidation appears to be ongoing. For instance, the city of Tarapoto has shown high population growth in recent years and other cities in the region have expanded in similar fashion.

Pressure on the land and forest conversion

Before 1940, despite the existence of large agricultural production estates, significant areas of the Huallaga and Mayo valleys in San Martín displayed abundant forest cover. According to APECO (1995), annual deforestation had not surpassed 8 thousand ha per year for the period 1940–1960. During the period 1960–1975, total deforestation increased from 296,019 ha to 1,046,675 ha. Annual deforestation jumped to an average of approximately 50,057 ha per year.

During the period between 1975 and 1979, the region lost 171 thousand ha of natural forest, maintaining the same deforestation rate as the previous period. In 1979 the annual rate of deforestation declined slightly, but it was followed again by a marked increase. In 1983 the total deforestation reached 1,386,214 ha, resulting in a four-year loss of approximately 168 thousand ha. As of 1983, the deforestation of new areas of natural forests stopped. Moreover, some recovery of previously deforested areas took place. The substitution of pastures by rice and corn cultivation could explain this, in part because of the adoption of irrigation technologies. As a result, cultivated land increased during this period but not at the expense of tropical forests. Between 1987 and 1989 deforestation again increased coinciding with the peak of the drug trade in the region. According to APECO (1995), the rate of deforestation began to decrease again starting in 1990.

Drawing upon information from the national censuses of 1940, 1962, 1972, 1981, and 1993, in addition to information provided by APECO (1995) on accumulated deforestation for the period 1960–1993, we constructed a time series data of population, migrant population, and cumulative deforestation. The time series data allowed the construction of linear, logarithmic, and polynomial regression models. The results of these regression analyses are presented in Figures 3.3 and 3.4. The models indicate that during the analyzed period there is a strong association (R2 = 0,991) between the volume of population total and deforested area in the department. In addition this association is positive, thus a further increase of the population can be expected to lead to further tropical forest conversion.

Similarly, there is a positive association (R2 = 0,738) between the magnitude of immigrant population and the area deforested. Although this relation is weaker in comparison to that of total population, it strongly suggests that increased migratory flows will also lead to accelerated deforestation. In both cases, the regression model that better fits the data is a polynomial equation of the third degree. This implies that relation is stronger for low numbers of population and immigrants. When the numbers increase, the association becomes weaker.

Based on these regression models, it is possible to formulate the following hypotheses:

- Once a certain amount of pressure is placed upon forests, additional conversion into agricultural lands or other non-forest land uses becomes very costly. This results in the intensification of existing agriculture and an increase in productivity.
- The advantages of the San Martín region for agricultural production have a limit that has already been reached. This implies that limited additional population increase from new immigration or other sources would not necessarily increase pressure on the land.
- The San Martín region is undergoing a process of urbanization with a correlated expansion of urban activities. This will eventually result in increased levels of rural–urban migration.

Migration and deforestation caused by the Federico Basadre road

Construction of the Federico Basadre road began in the 1940s and was completed in the 1970s. This road connects the city of Pucallpa, located at the Ucayali River, with the city of Lima. During the construction of this highway, a very dynamic process of population growth and occupation was triggered, of which the most salient outcome is the growth of the city of Pucallpa. This city was transformed from a simple Amazonian river village into the second largest Amazonian city and the fourth largest city in all of Perú. Another marked change resulting from the construction of the road was

Figure 3.3. Regression estimates between the area deforested and the total population, department of San Martín

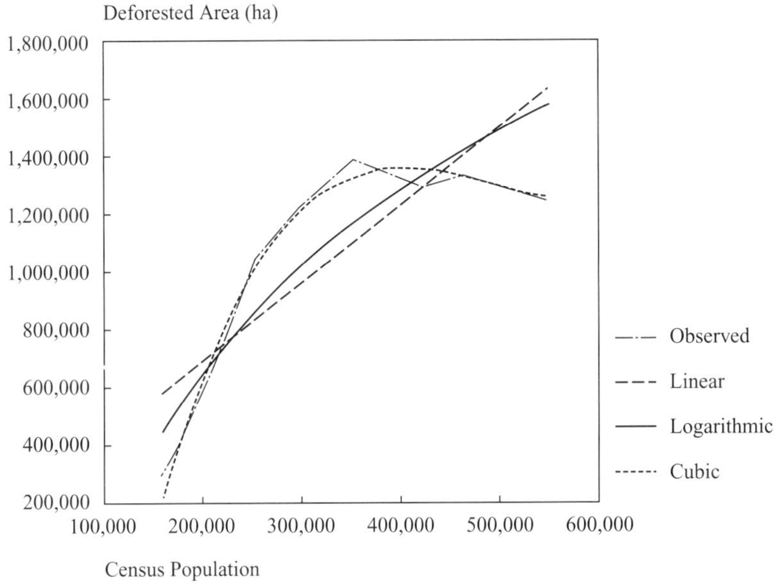

Figure 3.4. Regression estimates between area deforested and immigrant population, department of San Martín

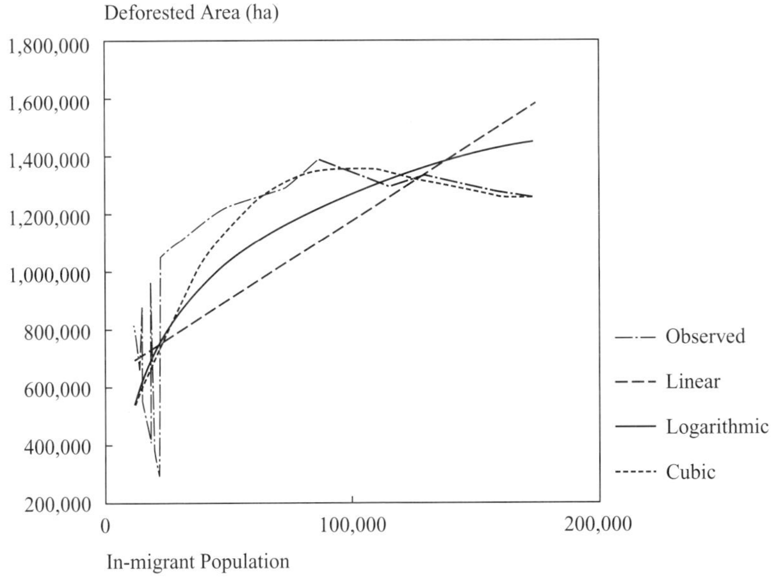

the shift in population concentration. Until 1940, the population of Pucallpa was centered on the Ucayali River and its tributaries, but since completion of the road, areas of high population density in Pucallpa have shifted to the areas surrounding the Federico Basadre road and its branches.

The population in areas influenced by this highway grew to 354,340 inhabitants, amounting to 82.3 percent of the total population of the department and more than 10 percent of the population of the Peruvian Amazon (Figure 3.5); 74.3 percent of this population lives in urban areas. Urbanization along the Federico Basadre road has increased in recent times, reflected in the growth of cities alongside this road, such as Pucallpa, Puerto Callao, Aguaytia, and others. Since the gradual construction of the road, the department of Ucayali has experienced demographic growth without precedent, reaching an annual growth rate of 5.6 percent during the last inter-census period. This growth rate is considerably higher than the national average during the same period and the second highest departmental growth rate after the department of Madre de Dios.

The rate of population growth in the zone can be attributed to massive migration into the region in the last few decades. This in-migration had already begun by the time the road was completed in the 1940s. Later, in the 1970s, agrarian reform prompted several farmers associations to established cattle production along this road. Initially, this attracted more immigrants, but it also consolidated the occupation of lands adjacent to the road, essentially prohibiting further settlement. More recently during the 1980s and the beginnings of the 1990s, the height of the drug trade resulted in the conversion of large areas to the cultivation of coca leaves. This generated economic prosperity in the zone, attracting many new immigrants. In addition, the social violence perpetrated by guerilla groups drove many farmers away from the Andes. Many of these farmers migrated to the city of Pucallpa and other areas of the region.

Figure 3.5 illustrates the trends and magnitude of the demographic explosion of the department of Ucayali. The population now living along the Federico Basadre road is largely a product of this explosion. The provincial and district capitals, located along other parts of the road network stemming from Federico Basadre, are also centers of particularly concentrated populations. These towns comprise around 12 percent of the population of the study area. The rest of the population (10 percent) lives in smaller villages in the region.

The impacts of human occupation along the road are evident in the deforestation of 507,077 ha (Figure 3.6) or 28 percent of this area (IIAP 1996; GOREU/IIAP 2003). This deforested area now primarily consists of

Figure 3.5. Inter-census population growth in the department of Ucayali between 1940, 1961, 1972, 1981 and 1993

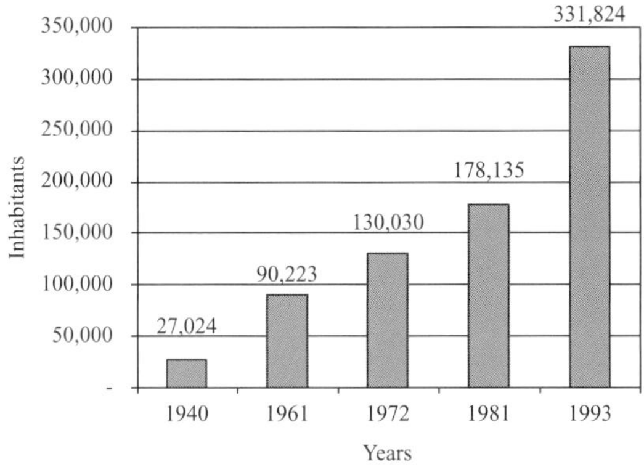

Source: INEI–National Censuses.

Figure 3.6. Accumulative deforestation in the department of Ucayali (1955–1998)

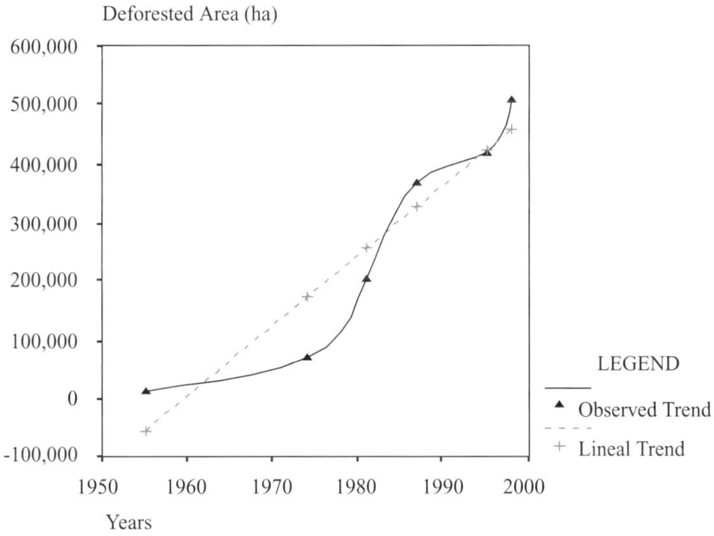

small agricultural swidden farms, pastures, and complexes of secondary forest. By reconstructing the data four relatively distinct periods can be distinguished:

1. The first period includes the completion of the Federico Basadre highway at the end of the 1970s and is characterized by a moderate but continued expansion of agricultural activities and moderate rates of annual deforestation.
2. The second period from the early 1980s until 1989 is distinguished by an accelerated expansion of agricultural activities. This stage was influenced by national policies promoting agricultural expansion through granting low cost credit loans. In addition, the expansion of drug trafficking and related coca leaf production led to the conversion of extensive wooded areas, particularly in the hilly parts of the region.
3. The third stage includes the years between 1990 and 1995. This period was typified by large-scale abandonment of pasture lands as a result of the activities of guerilla groups. During this period, the government had abandoned its policy of cheap agricultural credit as a direct result of the liquidation of the Agrarian Bank. Deforestation was moderate during this period and largely resulted from coca leaf production.
4. The final period from 1995 to 1998 was characterized by the gradual recovery of agricultural activities and, as a result, the opening of new agricultural land in forested areas. Deforestation rates increased during this period.

Intra-departmental migration and forest loss: The Iquitos–Nauta road

The 95 km Iquitos–Nauta road connects the cities of Iquitos and Nauta in the department of Loreto in the middle of the Peruvian Amazon lowlands. The idea of the Iquitos–Nauta road was conceived around 1885 as a direct result of a change in the flow of the Marañón River that made access to the port of Nauta more difficult. Various sections of the road were planned and constructed beginning in 1943. Construction continued periodically throughout the 1960s through to the 1980s, and small parts of the road were completed (OTAE 2001). However, only from the 1980s did the project proceed with any sense of direction. Finally, in 2005, the road was fully paved, completed, and officially opened.

After the rubber boom, when only a very small section of this road was completed, in-migration to the region was low and, as a result, little land was occupied and deforested. By the end of the 1960s, a number of larger estates producing sugarcane, rum, and other agricultural products, took over the few areas that were accessible from Iquitos. Migration until this time was rather marginal. From 1960 to 1980, several independent agriculturists settled in villages, taking advantage of trails that connected with Iquitos, which allowed limited commercialization of agricultural products. This occupation corresponded with the growth-dynamics of the city of Iquitos and its increased demand for agricultural products, but even this settlement was extremely limited (Wahl *et al.* 2003).

Significant migration to the area occurred simultaneously with the final and definitive outline of the road during the 1980s, and its subsequent construction in sections. Some of the migrations occurred spontaneously, while others were organized by agricultural associations. In the 1980s, several land distribution initiatives took place leading to the progressive occupation of the roadside land. These resulted in settlements such as Paujil, Nuevo Horizonte, Ex-Petroleros, and El Triunfo. In the space of a few years, the number of people receiving land grew, while at the same time unpaved access roads advanced. Soon there was much pressure to obtain ownership certificates which allowed access to loans from the Agrarian Bank. The impact of these two processes on the forests along this road was enormous. They led to the large-scale conversion of forests to agricultural production that in the end became unsustainable.

In the 1990s, new establishments were created and those that had already been established were consolidated. Many people who applied for land, and converted the forest on this land, eventually abandoned their holdings because they failed to develop profitable economic activities there. In 1981 the population along the Iquitos–Nauta road was around 2,500; by the year 2000 this population multiplied 6.5 times. Until the 1993 census, the annual population growth rate was 7.4 percent. This was very high in comparison with the regional growth rate of 3 percent.

Since 1996 the population growth rate along the road has increased in the order of 11 percent per year. Although the migration registration is unreliable for this period, we have been able to infer, based on a comparison of the rates of natural and real growth of the population since 1981 that the population has increased in the area in large part due to migration. As illustrated in Figure 3.7, in 1993, 44.5 percent of the population consisted of people that had migrated after 1981. This proportion increased to 49.1 percent in 1996 and to 66.6 percent in the year 2000. The population located along the Iquitos–Nauta road and its tributaries are mostly Amazonian mestizos. This means that most migrants come from other areas of Loreto (76.3 percent). Only 6.1 percent came from San Martín, 6.8 percent from Ucayali, and 9.4 percent from other departments of the Amazon. Migrants from the coast and the Andes represent less than 2 percent of the Iquitos–Nauta population (Figure 3.8).

There are some clear differences between this case and other colonization cases such as the Federico Basadre road, San Martín, Madre de Dios, and

Figure 3.7. Population growth along the Inquitos–Nauta road

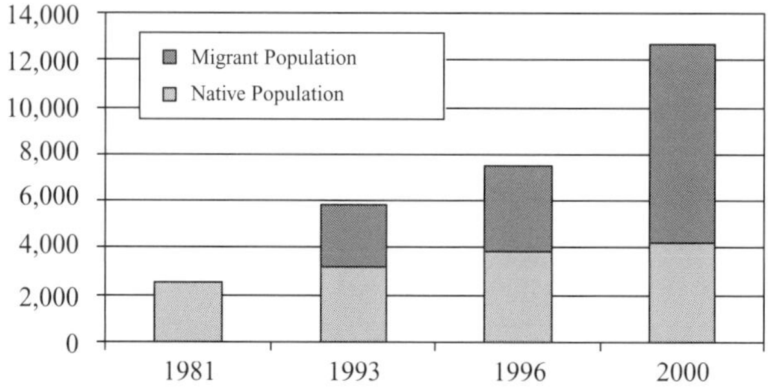

Source: INEI (1993); CTARL-IIAP (1996); Y AECI-IIAP (2001).

Figure 3.8. Place of birth of the population over 15 years of age along the Iquitos–Nauta road

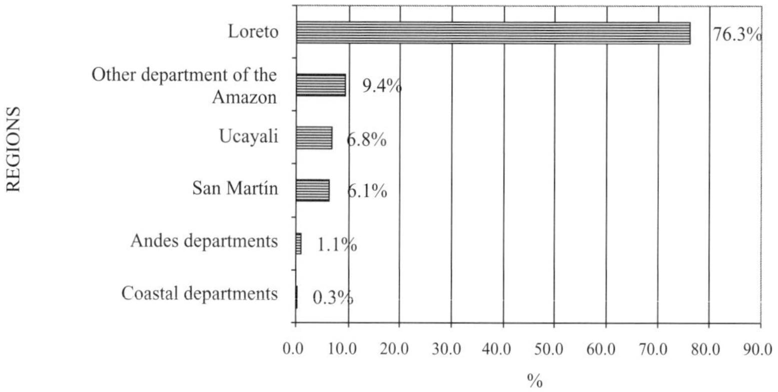

others. While the other examples feature migration largely related to Andean populations, the Iquitos–Nauta population has grown almost entirely because of immigration from Loreto-based people living along the major rivers in the region. Some of the migrants came from the neighboring city of Iquitos. Hence, this example implies a defined rural–rural and urban–rural migration trend. Interestingly, rural–rural and rural–urban migrations are occurring simultaneously; the population of Nauta is also increasing rapidly as rural settlers are attracted to new market opportunities created by the advancing road connection. Excluding those people who live along the Iquitos–Nauta

Figure 3.9. Evolution of deforestation along the Iquitos–Nauta road, 1987–2000

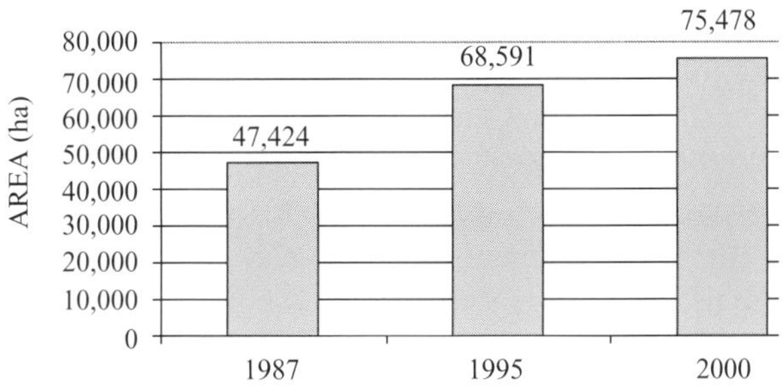

Source: IIAP *et al.* (2001).

road in the cities themselves, the road is home to some 12 thousand inhabitants, distributed in or near forty dispersed villages. If we considered the area of influence of this road, which spans from the west margin of the Amazon River to the east margin of Nanay river, the population of the area is approximately 42 thousand.

A diverse array of studies (IIAP *et al.* 2001; CTAR-RL 1997; CTAR-RL/IIAP 1996; Dourojeanni 1990; Gómez and Ortiz 1998) report on the suitability and capacity of this area for various uses. These studies agree that more than half of the area could be used for forest activities; 20 percent are vulnerable habitats that should be reserved for protection only; and just 20 percent could support agriculture or cattle production. IIAP *et al.* (2001) report that until 1987 some 47,424 ha of forests were converted to agricultural lands. This number increased to 68,591 ha in 1995 and 75,478 in 2000 (Figure 3.9). These areas are now small farms, areas of secondary forest, and remnant forests. Already, some of these areas are totally degraded and have no vegetation cover.

Conclusions

Our analysis of migration related to deforestation in the Peruvian Amazon, specifically in the three cases discussed in this paper, yields some useful conclusions. Processes of migration in the Amazon region are varied. Migration into the region alternates with migration out of the region. Both processes can be explained by several factors. In the Peruvian Amazon, factors include the emergence of new economic opportunities linked to booms of certain commercial forest products or agricultural crops. Migration is also linked to a broad range of government policies, some of which have been initiated to directly influence migration, while others affected migration towards the Amazon region indirectly.

Beside the pull-factors that encouraged people to move into the Amazon, push-factors propelled people away from their homes, particularly when migration from the Andes toward the Amazon became pronounced from the early 1970s. The construction of roads also played a vital role in several of the migration processes described here. It is not always clear whether the construction of roads can be labeled as pull- or push-factors.

In the example of San Martín, it was possible to make a clear quantitative link between migration and deforestation. However, the relationships between the other factors mentioned in this chapter that influence migration are far more difficult, if not impossible, to quantify. It is possible, nevertheless, to identify policies at national and regional levels that can have a positive or negative effect on migration patterns as we have shown above. It is also possible to discern the impacts that certain migration patterns may have on tropical forests at the destinations of migrants.

Our discussion has also revealed that there may be potential for designing policies that encourage migrants to use land in such a way that does not result in devastating deforestation rates. The Peruvian Government, like most governments in the Amazon basin, is not averse to forest conservation; they certainly prioritize, however, the economic development of the large groups of poor in the country. Initiatives that allow people from poorer regions to migrate to regions where they may have the possibility of a better life will certainly be attractive to Amazon countries. Some of the conclusions drawn here may provide an initial understanding of how measures to promote economic development-related migration can be complemented with measures that may mitigate the negative impact of this phenomenon on tropical forests.

Acknowledgments

We would like to thank the Japan Center for Area Studies for the invitation to present this paper at the symposium: "In, Out, In, Out: Populations, Migrations, and the Social Ecology of Tropical Forests". We also thank Deanna Donovan, Carla Staver and Karen de Jong Zumaeta for helping with editing the manuscript.

References

APECO. 1995. *Bases para la Gestión de los Recursos Naturales y Elaboración e un Plan de Ordenamiento Territorial de la Región San Martín*. Lima: Asociación Perúana para la Conservación de la Naturaleza (APECO).

APODESA. 1989. *El Modelo de Áreas de Producción (APM) y su Aplicación en la Selva Perúana*. Lima: Proyecto Apoyo a la Política de Desarrollo Regional (APODESA).

Aramburú, C.E. 1982. Expansión de la Frontera agraria y demográfica de la Selva Alta Perúana. In *C. Centro de Investigación y Promoción Amazónica, Colonización de la Amazonía*, 1–40. Lima.

Bilsborrow, R. 2002. Migration, population change and the rural environment. *ECSP Report*, 8: 69–94.

CTAR-RL. 1997. *Plan Estratégico para el Desarrollo Sostenible del Área de Influencia de la Carretera Iquitos-Nauta*. Iquitos: CTAR-RL/Comisión Multisectorial.

CTAR-RL / IIAP. 1996. *Zonificación Ecológica Económica del Área de Influencia de la Carretera Iquitos–Nauta (Estudio preliminar)*. Iquitos: Instituto de Investigaciones de la Amazonía Perúana -CTAR-L.

Dourojeanni, M. 1990. *¿Amazonía - Que Hacer?* Iquitos: CETA.

Gómez Romero, E. and T. Tamariz Ortiz. 1998. Uso de la tierra y patrones de deforestación en la zona de Iquitos. In C. Kalliola, R. and S. Flores Paitán, eds., *Geoecología y Desarrollo Amazónico: Estudio Integrado en la Zona de Iquitos, Perú*, 369–387. Annales: Universitatis Turkuensis Ser A II.

GOREU/IIAP. 2003. *Propuesta de Zonificación Ecológica Económica de la Cuenca del río Aguaytía*. Pucallpa: Gobierno Regional de Ucayali and Instituto de Investigaciones de la Amazonía Perúana.

IIAP. 1996. *Deforestación en el Área de Influencia de la Carretera Federico Basadre*. Iquitos: Instituto de Investigaciones de la Amazonía Perúana.

IIAP, AECI, and CTAR-RL. 2001. *Iquitos-Nauta: Zonificación Ecológico-Económica para el Desarrollo Sostenible. Publicación Digital en Disco Compacto*. Iquitos: Instituto de Investigaciones de la Amazonía Perúana, Programa de Ordenamiento Ambiental.

INEI (Instituto Nacional de Estadística e Informática). 1993. C*enso Nacionales 1993, IX de Población y IV de Vivienda*. Lima.

INEI. 1995. *Análisis Socio-Demográfico de la Región Ucayali*. Lima.

INEI. 2002. *Indicadores Demográficos*. www.inei.gob.pe. Lima.

ITDG. 1993. *Shapaja: Diagnóstico y Plan de Desarrollo Integral*. Lima: Intermediate Technology Development Group.

Limachi, L. 1995. *Proceso Migratorio en la Amazonía Perúana: Una Aproximación Preliminar*. Unpublished Manuscript.

Limachi, L. 2002. *Notas para Zonificación Ecológica Económica de la Región San Martín: Medio Socioeconómico*. Unpublished Manuscript.

Martínez, H. 1969. *Las Migraciones Internas en el Perú*. Lima: Monte Ávila editores.

OTAE. 2001. *Plan Estratégico para el Desarrollo Sostenible del Área de Influencia de la Carretera Iquitos–Nauta*. Iquitos: Borrador de trabajo.

Perz, S., C. Aramburú, and J. Bremner. 2002. *Population and Deforestation in the Pan Amazon Basin: A Comparison of Brazil, Bolivia, Colombia, Ecuador, Perú and Venezuela*. Manuscript, Department of Sociology, University of Florida.

Rodriguez, M. 1986. Poblamiento de la Amazonía desde el siglo XIX hasta 1940. *Shupihui*, XI (37): 7–28.

Rodríguez, V. 2001. *Monitoreo de la Deforestación y uso de la Tierra en el Área de Influencia de la Carretera Iquitos–Nauta*. Iquitos: Tesis de grado, Universidad Nacional de la Amazonía Perúana.

Salazar, A. 1984. Situación actual de los proyectos especiales de selva. In C. Consejo Nacional de Población and Centro de Investigación y Promoción Amazónica, eds., *Población y Colonización en la Alta Amazonía Perúana*, 245–281. Lima.

San Roman, J. 1994. *Perfiles Historicas de la Amazonia Perúana*. Iquitos: CETA.

Verdera, F. 2000. *Causas del Agravamiento de la Pobreza en el Perú desde Fines de la Década de 1980*. Lima: IEP.

Wahl, L., L. Limachi, and J. Barletti. 2003. Del discurso oficial al caserío rural: El desarrollo de la carretera Iquitos–Nauta. In C.E. Aramburú, and E. Bedoya eds., *Amazonía: Procesos Demográficos y Ambientales*, 156–182. Lima: Centro de Investigación Económica y Social.

4
Forest Product Extraction, Migration and Deforestation in the Northern Bolivian Amazon

Víctor Eduardo Llanque Zonta

Introduction

Culturally diverse populations of forest dwellers whose livelihoods are largely based on forest resources live in tropical forest ecosystems such as the Amazon. These forests are also inhabited by an extraordinary variety of plant and animal species. Increasing uncontrolled tropical deforestation not only jeopardizes biodiversity, but also threatens the survival of local populations that depend directly upon this ecosystem. Policy makers, scientists, and the general public, are increasingly concerned about tropical deforestation and its negative consequences. We therefore need to understand the factors underlying tropical deforestation and its impacts, so as to ensure the continuity of this ecosystem for future generations.

In this chapter I not only look at the direct linkages between forest product extraction, migration, and deforestation, but also analyze other intervening forces that affect this relationship. A simplistic perception of this relationship is that remote forest areas provide opportunities for groups of people deprived of prospects elsewhere, motivating them to migrate into remote forest areas where their activities impact upon the forest condition. A closer look at specific cases, however, often reveals a far more complex picture necessitating analysis in order to develop an understanding of the link between migration and forest condition. In an attempt to contribute to the understanding of the complex causes of increasing deforestation in the Amazon, I analyze the relationship between forest product extraction, migratory processes, and the impact of people's activities on the forest in the northern Bolivian Amazon—a region encompassing the whole Pando Department[1] and the Vaca Diez Province in Beni Department (Figure 4.1).

The chapter is divided into six sections. In section one I present the theoretical framework that enables analysis of the population, migration, and

Figure 4.1. Satellite image of the northern Bolivian Amazon

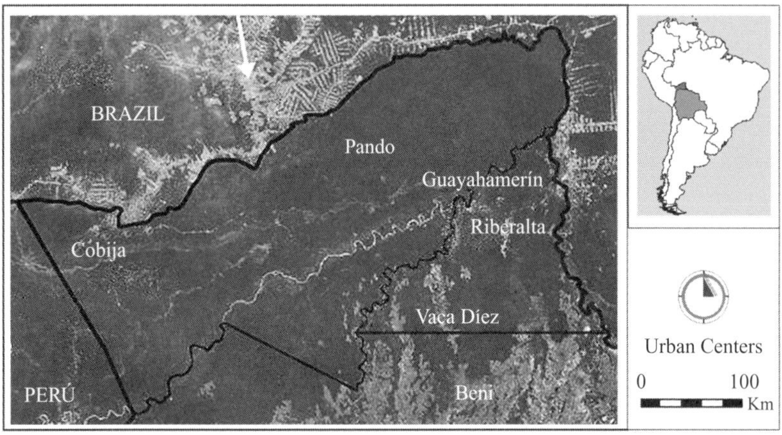

deforestation linkages. In the second section I trace the economic history of northern Bolivia, which is based on forest exploitation, and discuss how this has influenced the primary processes of migration. In the next section I analyze contemporary instances of migration into and within northern Bolivia. In section four I provide explanations of these modern migration processes, taking into account structural, "underlying" causes. In section five I analyze how migration has affected the condition of the forest in northern Bolivia. In section six I present the conclusions.

Theoretical framework

Push- and pull-factors are hypotheses basically used to explain the direct forces that influence people to migrate. This is based on the idea that the decision of whether to migrate or not is influenced by differences in wealth or development between the places of origin and destination (Carr 1997; Chrispin and Jegede 1998; Codrington 2002; Firebaugh *et al.*, cited in Stoian 2000: 212). Negative forces that may motivate migration are referred to as push-factors. The perceived advantages of the place of destination may become forces that attract migration. These forces are called pull-factors. In addition, it has been suggested that potential movements from one place to another are often influenced by the existence of alternative destinations and/or intervening opportunities (Stouffer 1940, cited in Chrispin and Jegede 1998: 26). These intervening factors may slow migrants down or may prevent them from migrating to their initially intended place of destination.

A large number of studies concerning the causes of deforestation and its human dimensions have been undertaken. Potter (Chapter 8) summarizes the three schools of thought concerning causes of deforestation. The neo-Malthusian school considers population to be the driving force behind deforestation, arguing that population density and deforestation are positively correlated. Alternatively, the two other schools, neo-classicism and political ecology, view economic principles as key factors underlying processes of deforestation. The former school emphasizes the role of the market and profit maximization in providing incentives for people to deforest. The latter blames powerful capitalists who claim large land areas for cultivation and cattle farming and invest in forest conversion for rent-seeking purposes.

Sunderlin and Resosudarmo (1999) propose an explanatory framework that facilitates the linkage of processes of deforestation and migration, a more difficult task when following the approach of Potter (Chapter 8). Sunderlin and Resosudarmo claim that it is necessary to distinguish three sets or levels of variables in order to understand more clearly the causes of deforestation. First, it is necessary to identify the agents of deforestation—the individuals or companies involved in land-use change, such as small farmers, loggers, and ranchers, whose decisions and actions are the direct sources of deforestation. These in turn respond to what Sunderlin and Resosudarmo (1999) call the agents' decision parameters or immediate causes of deforestation—external factors that influence the agents' decisions. Finally, they propose a set of underlying causes including broader economic, political, cultural, demographic, and technological forces that determine the immediate causes of deforestation. In this framework, immediate causes of deforestation can be factors that drive agents to migrate to forest areas. However, underlying causes may be triggering these direct factors.

Pacheco (1998) follows a similar framework to explain the causes of deforestation in Bolivia. However, he adds that forest conversion is not simply an effect of social or demographic tendencies or a result of economic development itself, but is instead an historical consequence of certain development styles. By development style he means the type of development process followed by a particular country or region. In this sense, development styles employed in Latin American countries have a common characteristic—they benefit from economic accumulation based on unrestricted use of natural resources. Moreover, the economic structures of these countries have resulted in an economy largely controlled by national or regional elites, while significant portions of the population are excluded from these benefits. These processes create considerable inequalities in the distribution of forestry incomes (CDMAALC, cited in Pacheco 1998: 32).

The universality of the principles mentioned above have been suggested by Kummer (1992), who proposes a similar course of events regarding the underlying causes of migration and deforestation in the Philippines. He argues that migration is caused by a lack of opportunities in places where unemployment and poverty are widespread, epitomizing the social inequity characteristic of the entire Philippine state.

Forest product extraction and the history of migration

Dispersed indigenous groups originally inhabited northern Bolivia, subsisting on hunting, fishing, and forest product extraction (CIDOB, cited in Stoian 2000: 51). These early inhabitants regularly crossed the borders to Perú and Brazil, showing a high level of geographical mobility, but their impact on the forest was practically negligible (Lockhart and Schwartz, cited in Stoian 2000). Larger migrant populations came into the region to exploit forest products as early as the nineteenth century. Initially, migrants came to extract Peruvian bark (*Cinchona spp.*) to make quinine, a medicine used to treat malaria. The exploitation of rubber that had its basin-wide boom between 1880 and 1920 brought large inflows of populations to northern Bolivia (see Limachi *et al.*, Chapter 3).

The emergence of Asian rubber in international markets had a profound effect on labor arrangements in northern Bolivia, leading to a significant process of intra-regional migration. Prior to this, the large rubber entrepreneurs held customary ownership over sizable estates, called *barracas*. These included extensive areas of forests and also central and satellite rubber accumulation points. The workforce that collected rubber lived in the central accumulation point that contained houses, stores, and rubber deposits. After Asian rubber entered international markets, northern Bolivia shifted to the collection of brazil nut (*Bertholetia excelsa*), combined with the ongoing extraction of rubber. Many of the workers who had lived in *barracas* relocated to new independent communities. Some of these were located on the same sites as the *barracas*, but others were established in locales more suited to the needs of the new inhabitants.

Rubber extraction surged again during World War II, when Asian sources were unavailable to Allied Forces. The rubber crisis after the war led again to an intensification of intra-regional migration through the reorganization of forest product extraction. The slump in rubber exports from Bolivia in combination with the Agrarian Reform of 1952 provided renewed incentives for migration out of *barracas* and into the independent communities. Table

Table 4.1. Deforestation in the northern Bolivian Amazon region, Bolivian Tropical Lowlands, 1980s–1990s, and Bolivia, 1970s–1990s

Region	Land area deforested[a] (%)		Average annual deforested[b] (%)	Land area (km^2)	Original forest area[c] (km^2)	Remaining forest area (km^2)	Total deforested area (km^2)		Annual rate of deforestation[d] (km^2/year)
	Time 1	Time 2					Time 1	Time 2	
Northern Bolivian Amazon Region	**1985**	**1990**	**1985–1990**				**1985**	**1990**	**1985–1990**
Northern Beni (Riberalta and Guayaramerín)	1.0	1.6	0.13	19,088	18,325	18,036	181	289	23
Pando	2.5	2.9	0.08	63,830	63,110	61,280	1,600	1,830	49
Overall	2.2	2.6	0.09	82,918	81,435	79,316	1,781	2,119	72
Bolivian Tropical Lowlands	**1985**	**1990**	**1985–1990**				**1985**	**1990**	**1985–1990**
Beni	1.9	2.4	0.12	196,270	85,490	83,410	1,610	2,080	100
Pando	2.5	2.9	0.08	63,830	63,110	61,280	1,600	1,830	49
Santa Cruz	6.7	7.6	0.19	224,690	180,240	166,480	12,160	13,760	340
Overall	4.6	5.3	0.15	484,790	330,670	311,170	15,370	17,670	489
Bolivia	**1975**	**1993**	**1975–1993**				**1975**	**1993**	**1975–1993**
Overall	nd	5.4	0.30	1,098,581	564,684	534,442	nd	30,242	1,680

Sources: Bolivian Overall: GEOBOL (1978) and MDSMA (1995) in Pacheco (1998: 53, 56); Bolivian Tropical Lowlands: CUMAT (1992) in Pacheco (1998: 57); Northern Beni deforestation: ERTS/GEOBOL (1978) and MDSP (1999) in Usipales (2001: 58).

Notes

a: Percent deforested refers to the deforested land area as a percentage of original forest area, as of the year stated.

b: Average annual deforestation refers to net forest loss per year as a percentage of original forest area.

c: Original forest area refers to the addition of remaining forest area plus the total deforested area at time 2.

d: Annual rate of deforestation refers to the average of the deforested area during the period time 1 to time 2.

4.1 provides official information on demographic data to illustrate internal population mobility in northern Bolivia between 1976 and 2001.

Modern migrations

Immigration to northern Bolivia

To date, northern Bolivia is not very densely populated in comparison to other regions of the country—only some 169 thousand inhabitants live in the department of Pando and the Vaca Diez Province. This results in a density of 2.04 inhabitants per km^2, compared to the national figure of 7.6 inhabitants per km^2 (Table 4.2). More than 70 percent of the regional population now lives in just three urban centers, Riberalta (54 percent), Guayaramerín (27 percent), and Cobija (17 percent). Until the end of the 1990s, the number of immigrants into the Bolivian lowlands was relatively low, however, the migration percentages for this region are quite significant (Table 4.3). The immigrant population in this area increased by 8.1 percent per year until 1992.

Migration inflows from regions outside northern Bolivia were not very significant in rural areas, but were higher for urban centers. The rural population has been growing primarily due to internal growth rate, suggesting insignificant extra-regional immigration into these areas. Migrants from other regions that come to cultivate or live in rural areas are forced to settle in the colonization areas assigned by the government (High Beni, La Paz, and Santa Cruz), whereas in northern Bolivia, frontier development relies upon private sector endeavor despite the fact that no major colonization efforts have been initiated since the rubber boom (Stoian 2000). The majority of extra-regional immigrants settle in the urban centers of this region. This migration pattern has been particularly intense after the opening of the road that connected northern Bolivia with the rest of the country in 1986, and there has been a recent push of emigration to the city of Cobija, capital of Pando, that is located directly at the Brazilian border. Traders or merchants from the Andes come into the region to sell their products in the local markets and to benefit from direct access to the Brazilian frontier. This is reflected in the high percentage of new immigrants for this region between 1987 and 1992, immediately after the opening of the roads (Table 4.3). In Vaca Diez Province, these populations settled in Riberalta and Guayaramerín. The former is a large market itself, attracting merchant trade. The latter is on the Brazilian border and is thus the channel for importing industrialized products from that country.

Table 4.2. Area, population, urbanization, density, and area per inhabitant in northern Bolivia, the Bolivian Lowlands and Bolivia

Place	Total land area (ha)	Population (2001) Rural	Urban	Total	Urban (%)	Population density (inhab/km²)	Area per inhabitant (ha/inhab)
Northern Bolivia Amazon							
Northern Department of Beni							
Riberalta	1,271,728	11,466	64,511	75,977	84.91	5.97	16.74
Puerto Guayaramerín	637,084	7,349	33,095	40,444	81.83	6.35	15.75
Overall northern Department of Beni	1,908,812	18,815	97,606	116,421	83.84	6.1	16.4
Department of Pando							
Cobija (Capital)	44,746	1,504	20,820	22,324	93.26	49.89	2.00
Porvenir	106,465	3,713	-	3,713	-	3.49	28.67
Bolpebra	248,406	1,194	-	1,194	-	0.48	208.05
Bella Flor	523,186	2,305	-	2,305	-	0.44	226.98
Puerto Rico	524,180	4,003	-	4,003	-	0.76	130.95
San Pedro	263,062	1,082	-	1,082	-	0.41	243.13
Filadelfia	1,182,721	3,145	-	3,145	-	0.27	376.06
Puerto Gonzalo Moreno	127,404	3,810	-	3,810	-	2.99	33.44
San Lorenzo	319,807	3,471	-	3,471	-	1.09	92.14
Sena	757,351	2,240	-	2,240	-	0.3	338.10
Santa Rosa del Abuná	438,380	2,097	-	2,097	-	0.48	209.05
Ingavi (Humaita)	514,921	899	-	899	-	0.17	572.77
Nueva Esperanza	362,207	740	-	740	-	0.2	489.47
Villa Nueva (Loma Alta)	281,695	993	-	993	-	0.35	283.68
Santos Mercado	679,327	509	-	509	-	0.07	1,334.63
Overall Department of Pando	6,373,858	31,705	20,820	52,525	39.64	0.82	121.35
Overall northern Bolivia Amazon	8,282,670	50,520	118,426	168,946	70.10	2.04	49.03
Beni overall	61,628,500	249,152	113,369	362,521	31.27	1.70	170.00
Santa Cruz	22,469,000	483,823	1,545,648	2,029,471	76.16	9.03	11.07
Tropical Lowlands overall	68,486,181	1,826,505	618,012	2,444,517	25.28	3.57	28.02
Bolivia	10,944,843	3,108,443	5,165,882	8,274,325	62.43	7.56	1.32

Source: Based on INE (2001).

Table 4.3. Overall inter-state migrants, plus rural and urban in-migration in the Tropical Bolivian Lowlands, late 1980s–early 1990s

	Overall			Rural areas			Urban areas			
	Population age ≥5[a] (1992)	Migrants 5 years to 1992	Migrants 5 years to 1992 (%)	Population age ≥5 (1992)	Migrants 5 years to 1992[c]	Migrants 5 years to 1992[c] (%)	Population age ≥5 (1992)	Migrants 5 years to 1992[c]	Migrants 5 years to 1992[c] (%)	Migrants in urban areas (%)
Beni	225,343	19,344	8.6	73,403	4,281	5.8	151,846	15,078	9.9	78
Pando	31,094	4,897	15.7	22,557	3,152	14.0	8,540	1,753	20.5	36
Santa Cruz	1,130,140	88,181	7.8	306,156	17,015	5.6	823,949	71,681	8.7	81
Total	1,386,577	112,422	8.1	402,116	24,448	6.1	984,335	88,512	9.0	78

Source: INE (1992), in Perz (2002: 23–26).

Notes

a: The population observed is the population age ≥5.

b: The migrants observed here are the population age ≥5 that declared a place of residence 5 years prior to the census.

c: Migrations estimates here may not total those in the Overall figures, this is because some migrants did not declared whether they lived in rural or urban areas.

Intra-regional migration

Intra-regional migration provides a more complex picture. Brazil nut and rubber collection continued to be the main economic activities in northern Bolivia until the late 1980s. These activities provided year-round occupation for collectors. Until the mid-1980s, the Brazilian government paid a premium over nationally produced rubber to support Amazonian collectors. Bolivian collectors benefited from this as they sold most of their rubber to Brazilian merchants. The abrogation of these subsidies during the mid-1980s, in combination with Bolivia's hyperinflation, explains the decline in Bolivian rubber exports. Since then, forest product extraction has been largely limited to brazil nuts. Because forest product extraction ceased to provide a year-round source of income, forest dwellers saw no other alternative but to increasingly move to independent communities or urban centers in the region. A village-level survey undertaken by Stoian in rural rubber and brazil nut estates and independent communities revealed that between 1986 and 1997, 81 percent of the *barracas* had experienced out-migration, 16 percent had stable populations, and only 3 percent had experienced in-migration (2000). During the same period, 63 percent of the independent communities experienced in-migration, 31 percent emigration, and 6 percent had stable populations. Between 1992 and 1997, the *barracas* lost on average 34.6 percent of their population to out-migration, while the population in independent communities increased by 47.5 percent (adjusted to account for internal growth). Combining the migration losses and gains of *barracas* and independent communities, the population of the surveyed settlements grew by 10.7 percent in the period 1992–1997. Immigration was most intense into the peripheral-urban centers of Riberalta that grew by 84 percent during that period (Stoian 2000).

Photo 4.1. Dirt road that provides increased access to the forest of northern Bolivia, contributing to intra-regional migration

During this period, the department of Pando and the province of Vaca Diez experienced high rates of urbanization, especially after 1985. The town of Riberalta had the highest influx of people between 1980 and 1987 when its population grew by 6.7 percent annually. As a result, the year 1985 marks a clear turning point for the department of Pando. Previously, Pando's population had grown by 3.5 percent annually, but from 1985 to 1992, the overall annual growth rate went into the negative (-2.8 percent) as a result of pronounced out-migration, mainly from its rural areas. Furthermore, whilst the population of Pando's rural areas were declining at -6 percent, the urban areas in this department and Vaca Diez Province were growing annually by 10.9 percent and 6.4 percent respectively. These figures clearly show the intense rural–urban migration that occurred between these years. Urbanization, however, was apparent in Riberalta even before 1985;[2] between 1976 and 1985 the annual growth was already quite high (5 percent).

Another particularly intense migratory pattern in the region is that of circular (seasonal) migration from Riberalta in and out of *barracas* before and after the brazil nut harvest season. Currently, between 5,000 and 6,000 urban-based brazil nut gatherers (locally called *zafreros*), participate in the brazil nut harvest (the *zafra*) which takes place between November and February/March. During this time the *zafreros* move to the forest, remaining for three months at a time.

During the most recent period (1992–2001), the rural exodus has virtually ceased and the urbanization process has slowed down. The rural populations in Vaca Diez and Pando have increased by 3.7 percent and 1.4 percent respectively, whereas the annual rates of urban population growth decreased from 6.4 percent between 1985 and 1992 to 3.6 percent between 1992 and 2001 in Vaca Diez, and from 10.9 percent to 8.1 percent in Pando.

The immediate causes of intra-regional migration

Both push- and pull-factors explain rural–urban migration into Riberalta (Llanque 2003). The low employment opportunities after the mid-1980s rubber crisis were the main factors pushing forest dwellers out of rural settlements. The prospect of employment opportunities in Riberalta, as a result of the expansion of the brazil nut processing industry (Bojanic 2001), was the primary incentive that pulled the forest dwellers into this town. Cobija and Guayaramerín attracted rural–urban migrants who sought to benefit from the intensive border-trading. Similar types of push- and pull-factors explain rural–rural migration from *barracas* to independent communities. While the lack of employment in the *barracas* after the rubber crisis pushed rubber and brazil nut collectors out, the opportunity to combine both agricultural activities and seasonal brazil nut collection "pulled" the migrants into these settlements.

At the same time, services provided in the rural towns and independent communities are additional, but no less important, pull-factors. Immigrants to independent communities and regional towns are attracted by educational and health services. Opportunities for schooling and health care in *barracas* are extremely low. From the ex-rubber tappers sampled in a survey of peripheral-urban households, approximately 7.1 percent had no education at all and approximately 41 percent had not completed primary school (Llanque 2003).

The underlying causes of modern migration

The previous section described the immediate causes of migration. Within the framework proposed by Sunderlin and Resosudarno (1999), there are likely to be underlying causes that trigger these immediate causes. Macroeconomic and specific policies are located in this level of causality. Bolivia experienced three main generations of reforms in the 1950s, 1980s, and 1990s; these shape current conditions and can account for some country-wide trends of migration and deforestation.

The first change came in 1952, when Bolivian economic policies shifted from a focus on mineral export to a more social strategy that aimed to diversify the economic structure and reduce dependence upon imports. The Bolivian government stimulated agricultural production in the lowlands to replace food-product importation and to solve the demographic pressure felt in the Bolivian Andes (Pacheco 1998: 152). This process was accompanied by: Agrarian Reform endowing land for agriculture; road constructions; colonization programs in the tropical lowlands; and credit and other subsidies for agricultural production in the tropical lowlands. As a result of these measures, migration and deforestation gradually intensified, particularly into the eastern regions of the country with the nuclei of Cochabamba and Santa Cruz. Nevertheless, this process was not extensive since agricultural production was oriented toward the small-scale internal market. These changes did not have a particularly significant impact in northern Bolivia. None of these policies focused on developing the forestry sector, as this was not considered a priority within the strategy of product diversification. However, the agrarian reform stimulated internal migration in northern Bolivia as described previously.

In 1985, Bolivia adopted its New Economic Policy (NEP) aimed at stabilizing inflation. The NEP implemented the typical structural adjustments of a neo-liberal political system: the almost total liberalization of markets for goods, credit, capital, and work through the removal of the control of internal prices; the abolition of quantitative limits for international trade; and the reduction in public investment and subsidies. This new macro-economic model relied upon export production through a new and dynamic private sector (IDH 2002). The structural adjustments stimulated non-traditional exports through tax reduction and improvements in road communications and other facilities aiding export. These changes had a profound impact upon agricultural expansion and related migration patterns. They especially stimulated migration towards the eastern part of Bolivia, in particular the Santa Cruz region, and led to agricultural expansion and related deforestation (Pacheco and Mertens 2004). As a result, deforestation in the Bolivian tropical lowlands increased drastically after the 1980s.

During the 1990s, a third generation of reforms countered to some extent the consequences of the economic policies, promoting environmentally friendly and socioeconomic development. The establishment of the Ministry of Sustainable Development, the Ecological Pause that limited expansion of forestry activities, the Environmental Law (1992), the Popular Participation Law (1994), the New Agrarian reform (1996), and the Forestry Law (1997) are all part of this third generation of reforms. Between the 1950 and 1995 the percentage of land under cultivation in the tropics has increased from 15 percent to 58 percent of the national cultivated area (Figure 4.2). From this tropical cultivated area, approximately 64 percent is currently worked by companies that primarily produce soybean for export, while the remainder is used by *campesinos* who largely practiced subsistence agriculture.

Figure 4.2. Changes of proportion of cultivated land area of the two regions of Bolivia: Andes and Tropics, between 1950 and 1995

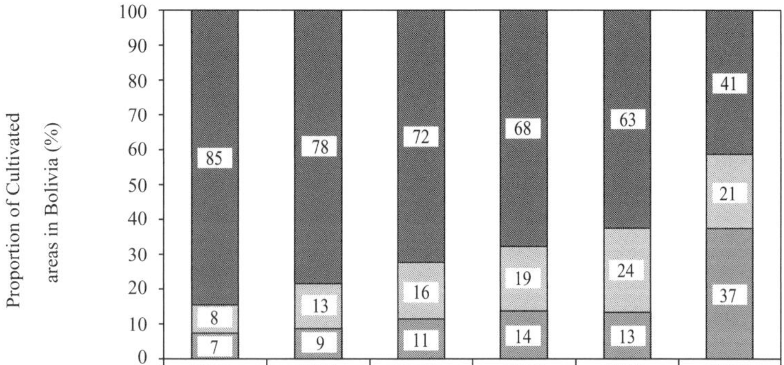

	1950	1964	1970	1979	1985	1995
Andean Agriculuture of Campesinos	85 (%)	78	72	68	63	41
Tropical Agriculture of Campesinos	8	13	16	19	24	21
Tropical Agriculture of Companies	7	9	11	14	13	37

The third generation policies discussed above have had little, yet noticeable, impact upon the northern Bolivian region. Aside from the effects of the agricultural reform of the 1950s, the road construction program during the 1980s has made northern Bolivia accessible, which to some extent explained the inflows of extra-regional immigrants—primarily merchants and traders—and the expansion of the brazil nut processing industry. The latter was one of the pull-factors that attracted intra-regional migrants to the major towns. The third generation of reforms has been having an impact upon northern Bolivia since the late 1990s. Land ownership and access to forests have been redefined through the changes resulting from the new forestry law and land reform legislation. It is highly likely that these changes may become a pull-factor sooner rather than later, attracting extra-regional immigrants to northern Bolivia once more.

Impacts of migration on the forest

The direct impact of migrants upon the forest in northern Bolivia has changed over time. Forest loss was more intense around *barracas* during the various rubber production periods. After the cessation of rubber production, pressures on the forest shifted to the new settlements where many former rubber tappers had relocated, adding subsistence agriculture to their portfolio of economic activities. In addition, a number of former *barraca* owners took up cattle ranching in their forest estates. Since the early 1990s, logging has also become increasingly significant in northern Bolivia (Bojanic 2001), and a new group of small-scale commercial farmers who supply the regional towns with foodstuff have entered the scene. Further, urban dwellers impact upon the forest, partly to obtain residential space and construction material, but also because of an increasing demand for local agricultural and forest products.

For the rubber tappers, the underlying motivation for cultivating food crops is subsistence, although commercial production is increasing. For this neglected group of people, deforestation is unavoidable if they want to survive. The amount of land under cultivation has increased over the past years: between 1980 and 1990, the area under cultivation in the departments of Beni and Pando increased from 18,000 to 33,000 ha and from 3,000 to 15,000 ha respectively (Henkemans 2001). Agriculture in this region is based on a system of shifting cultivation. Deforested plots (*chacos*) are sparsely distributed and left to rest for long periods after two years of use. The agricultural activities of this group are not very intensive and involve relatively minor forest conversion. The rates of forest conversion into arable land are higher close to the Brazilian border (Henkemans 2001). Large scale conversion of forestland for commercial purposes has little potential in this region due to low soil fertility.

The *barraca* owners are occupied with brazil nut extraction between November and March each year, bringing in seasonal collectors from the main towns in the region. For the rest of the year they engage in cattle ranching and timber extraction. Cattle ranching is most intensive in Pando. It is also concentrated along the new roads that connect Cobija with Riberalta, and Riberalta with Guayaramerín. The costly conversion of forest into pasture discourages cattle ranching in the natural savannas predominantly located south of the northern Bolivian Amazon region. The intensity of cattle ranching has significantly increased in recent times.[3] This is in part a result of the need to prove the economic value of forestland in order to obtain legally recognized land ownership.

Table 4.1 compiles information on deforestation in Bolivia, its tropical lowlands, and the northern Bolivian Amazon between the 1980s and 1990s. Between 1975 and 1993, the forested area in Bolivia decreased from 565

Photo 4.2. Conditions that favor cattle production and traditional forest users who leave the forest contribute to wide-scale forest conversion

thousand km^2 to about 534 thousand km^2 (Pacheco 1998). During that time, approximately 30,242 km^2 of the Bolivian forests was converted at a rate of 1,680 km^2 per year. Between 1985 and 1990, the highest annual rate of deforestation was seen in the department of Santa Cruz (approximately 340 km^2 per year), while the lowest was reported in the department of Pando (49 km^2 per year). The department of Beni lost only 1.6 percent of its forest during this period and the entire northern Bolivia lost just 2.6 percent (Usipales 2001). These data are likely to have changed during the 1990s as a result of, for example, the expansion of cattle ranching in Pando or land reform.

These figures demonstrate that the correspondence between forest loss, rural migration, and urban migration is moderate. The rightmost column in Table 4.3 shows the percentage of immigrants who have settled in urban areas. During the 1980s and 1990s, the rate of urbanization in the Bolivian lowlands has increased turning this region into an urban frontier. In northern Bolivia, the urbanization process implies that most migrants settle in urban centers, with perhaps only indirect pressures placed on the forest. However, when we compare national-level data, the highest percentage of immigrants settling in rural areas is apparent in Pando. This concurs with the highest rates of deforestation in Santa Cruz, which is clearly the most urbanized department.

Conclusions

Migration in northern Bolivia has had four main features: 1) migration of people from other areas into the region; 2) rural–urban migration within the region; 3) urban–rural migration within the region; and 4) rural–rural migration within the region. Different forms of migration have come into play at different times in history, and each of the four types delimited above can be linked to a relatively clear combination of forest- and non-forest-based activities. External macro-forces influenced these migration patterns. Migration into the region was of major importance when international demand for the region's forest products existed. International factors, such as the supply of rubber from Asia, World War II, and the abrogation of rubber subsidies in Brazil, influenced the feasibility of certain types of forest product exploitation and related contractual agreements, and this heavily influenced intra-regional migration.

The factors that shape migration in northern Bolivia are likely to be dynamic for years to come. Two key factors are the expansion of Cobija as the political center in the region and the planned bi-oceanic road connecting Brazil and Peru and passing within kilometers of northern Bolivia. This will create an entirely new set of conditions for northern Bolivia that the region has not yet experienced. Another factor is the redistribution of forestland among the rural indigenous population and *campesinos* that is now taking place. These macro-factors are easily identifiable. Localized factors are more difficult to descry. Early trends suggest that a large number of migrants will move closer to the border regions and towards Pando. In addition, there are likely to be new waves of extra-regional migrants into northern Bolivia. The impact of these new waves of migration remains difficult to foresee.

Although the particular pattern of migration described here is typical for the non-exhaustive extraction of forest products like rubber and brazil nut, similar complex patterns of multidirectional migration are common for tropical regions where local resource use is somehow linked to larger markets (see Limachi *et al.*, Chapter 3). The connection between migration patterns and deforestation in the case of northern Bolivia is noteworthy. Some of the major migration waves into the forest regions had little impact upon the forest, while subsequent rural–urban migration resulted in an increased pressure upon the forest. These variations are largely the result of shifts in the primary economic activities of migrants, as they change focus from forest product collection to commercial agricultural production. From this follows the conclusion that migration towards a forested region in itself cannot account for increased pressure on the forest. Interrelated factors that must be taken into account include the type of resource use associated with the particular migration trend, and on subsequent, more localized migration processes. Under certain circumstances, migration out of the forest may imply a shift in resource use that could result in increased pressure on the forest.

Acknowledgements

It is my pleasure to acknowledge Armelinda Zonta and Oscar Llanque for their great support; their advice and help have been crucial to the culmination of this paper. I also want to thank the Japan Center for Area Studies and Wil de Jong for encouragement and inviting me to present this paper.

Notes

1. The term “department” refers to the political geographic division of regions within Bolivia. “Province” refers to the political sub-division of territories within a “department”.

2. The negative population growth for Pando (-0.6 percent) was not unique in Bolivia. In the same periods, the rural areas of La Paz, Oruro and Potosi departments also experienced out-migration, with growth rates of -0.5 percent, -1.6 percent and -0.6 percent respectively (Banco Santa Cruz, in Stoian 2000). None of Bolivia's nine departments were exempted from the rural exodus, with the 1976–1992 net-migratory loss amounting to 3 percent of the rural population. The negative rates exhibited by rural Pando and Vaca Diez therefore should not be viewed solely as a response to the latest rubber crisis but are also related to the general rural–urban migratory trend in Bolivia.
3. Unfortunately, there is no data to show this expansion. The latest estimates are from 1990. This shows the need for new research in this area.

References

Bojanic, A.J. 2001. Balance is beautiful: Assessing sustainable development in the Bolivian Amazon. *PROMAB Scientific Series* 4. Programa de Manejo de Bosques de la Amazonia Boliviana (PROMAB). Utrecht: University of Utrecht.

Carr, M. 1997. *New Patterns: Process and Change in Human Geography*. United Kingdom: Thomas Nelson and Sons Ltd.

Chrispin, J. and F. Jegede. 1998. *Population, Resources and Development*. London: Collins.

Codrington, S. 2002. *Planet Geography*. Adelaide: Solid Star Press.

Henkemans, A.B. 2001. *Tranquilidad* and hardship in the forest: Livelihoods and perceptions of *Camba* forest dwellers in the Northern Bolivian Amazon. *PROMAB Scientific Series 5*. Programa de Manejo de Bosques de la Amazonia Boliviana (PROMAB). Utrecht: University of Utrecht.

IDH (Informe de Desarrollo Humano en Bolivia). 2002. *Programa de las Naciones Unidas para el Desarrollo*. La Paz.

INE (Instituto Nacional de Estadística). 2001. *Censo Nacional de Población y Vivienda 2001*. Beni: Resultados Departamentales. La Paz: INE.

INE. 2001. *Censo Nacional de Población y Vivienda 2001*. Pando: Resultados Departamentales. La Paz: INE.

IPHAE (Instituto para el Hombre, Agricultura y Ecología). 2003. *Departamento de Apoyo al Fortalecimiento Organizativo y Tenencia de la Tierra*. Riberalta: IPHAE.

Kummer, D.M. 1992. *Deforestation in the Postwar Philippines*. Research Paper 234. Chicago: Department of Geography, University of Chicago.

Llanque Zonta, V. 2003. *Driving Forces of Rural–Urban Migration in the Northern Bolivian Amazon Region*. Flekke: Red Cross Nordic United World College.

Pacheco, P. 1992. *Integración Económica y Fragmentación Social: El Intinerario de las Barracas en la Amazonía Boliviana*. La Paz: Centro de Estudios para el Desarrollo Laboral y Agrario (CEDLA).

Pacheco, P. 1998. *Estilos de Desarrollo, Deforestación y degradación de los Bosques en la Tierras Bajas de Bolivia*. La Paz: CIFOR/ CEDLA/ TIERRA.

Pacheco, P. and B. Mertens. 2004. Land use change and agriculture development in Santa Cruz, Bolivia. *Bois et Forets des Tropiques*, 280: 29–40.

Stoian, D. 2000. *Variations and Dynamics of Extractive Economies: the Rural-Urban Nexus of Non-timber Forest Use in the Bolivian Amazon*. PhD Thesis. Freiburg: Albert-Ludwigs-Universität.

Sunderlin, W.D. and I.A.P. Resosudarmo. 1999. The effect of population and migration on forest cover in Indonesia. *Journal of Environment and Development*, 8(2): 152–169.

Usipales, O. 2001. *Incidencia de la Ganadería en la Deforestación del Bosque Tropical de Tierra Firme en el Municipio de Riberalta*. Graduate Thesis. Riberalta: Universidad Técnica del Beni (UTB).

5
Migration and Forests: An Overview from Africa

Philip N. Bradley

Introduction

By global standards, Africa is relatively sparsely populated and extremely large. The total population for Sub-Saharan Africa is 643.3 million (1999 estimate: World Bank data[1]). Figure 5.1 shows schematically how it compares to some other sub-continents in terms of size and population. Even if the uninhabitable Sahara and Namibian deserts are withdrawn from the picture, there still remains a vast area of land for Africa's existing population to occupy. Africa's current population, however, is increasing rapidly: "In 1950, Africa's population was less than half that of Europe. But 50 years later, Africans easily outnumber Europeans. At a current growth rate of 2.4 per cent per annum, Africa will again double its population by 2035" (UNECA 2003: 1; see Table 5.1). Not only is Africa's population growing rapidly, it is also highly mobile. Both population growth and migration are key factors that influence natural resource use and affect Africa's tropical forests.

Concerns about forest status, forest resource use, deforestation, and the role of migrants are best addressed from a secure knowledge base. For both migration trends and forest conditions, but more particularly for the latter, the quality of our knowledge and of the narratives that surround it is questionable. There is a great deal of uncertainty about the true picture of African forests (see Fairhead and Leach 1998). Unfortunately, despite concerted efforts by a number of agencies over the last twenty years, the true status of Africa's rainforests remains largely unknown. This is particularly the case for deforestation rates and causes, complicating assessment of the impact of migration trends on the condition of Africa's forests.

In this chapter I first consider the demographic situation in Africa south of the Sahara, looking at the historical trends that underpin the current situation and casting an eye into the future. I adopt both a continental and a regional scale and, in some instances, pursue a more local focus—all in order to present a general overview of the current dynamics and processes that drive

Figure 5.1. Africa's size and population compared with other sub-continents. The shapes of China, the EU, the USA, India and Japan have been adjusted to fit within the outline of Africa, but their relative sizes are maintained. Figures show total populations (in millions)

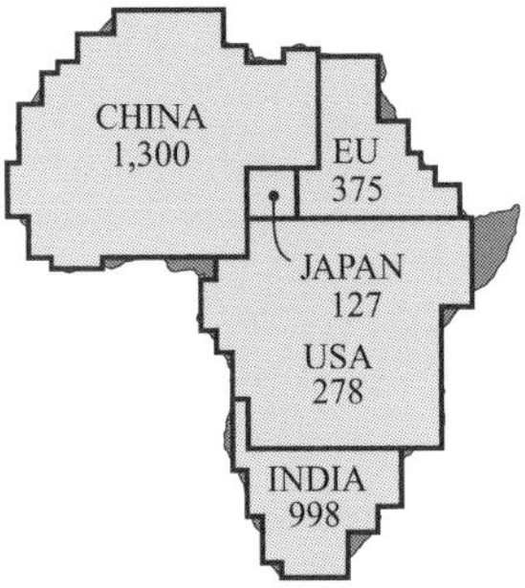

	Area (1000 km^2)	Population 1999 (m)	Population density (persons per km^2)
Africa (humid forest states)	6,297	283.8	45
South East Asia	4,361	551.4	127
Amazonia	6,860	26.8	4

Table 5.1. Sub-Saharan African population estimates

Year	millions
1500	47
1840	104/50[a]
1880	120
1900	129
1920	142
1935	165
late 1940s	200[a]
early 1960s	300[a]
1999	643.3

Sources: Caldwell (1985); Reader (1997).

a: Reader states that the population of Sub-Saharan Africa was only 50 million as late as 1850, following Manning's (Reader 1997: 437) modeling of the demographic impact of the slave trade.

migration. This is followed by a review of contemporary knowledge regarding the forests of Africa (in particular the humid forests of the equatorial zone) and the processes that determine their status.

In section two of this chapter I discuss the histories of African migrations. In section three I analyze key factors that influence contemporary migratory and demographic processes of Sub-Saharan Africa. In the following section I provide more detail while analyzing migration that is directly linked to particular forest resources. In section five I discuss various aspects of

Africa's forest cover and how it is changing, linking these transformations to migration. In the last section I present the conclusions.

Histories of movement

For many centuries, Africans have been on the move, seeking new land to farm or fleeing persecution. A useful starting point is the emergence of the slave trade in the early sixteenth century, which particularly affected the humid forest regions of West Africa and the Congo Basin. In both sub-regions, slave raiding had dramatic impacts on population and the movement of people, particularly along what came to be known as the slave coast of West Africa, stretching from present-day Nigeria to Senegal. The arrival in the late seventeenth century of flintlock guns altered the dynamics of warfare in Africa and escalated traditions of violence, including such activities as the raiding of the interior, the ravaging of villages and the depredation of people, the rise and fall of mini-empires and kingdoms, the constant hiding, and the relocation of settlements in the bush and forest, all of which have impacted upon natural vegetation and forests in particular (Dounias and Leclerc, Chapter 7).

Following the decline of the Atlantic slave trade came the rise of mercantilism, characterized by the emergence of plantation economies along the coast and in the forest zone. The Lever Brothers obtained a concession for oil-palm plantations in 1911 in the Belgian Congo, after having been refused permission in British West Africa[2]. Cocoa quickly became one of the principal cash crops of the forest zone. Unlike oil-palm plantations, cocoa could be commercially grown by local inhabitants on their own farms. This smallholder production system continues to thrive even today (Acheampong, Chapter 6). During this period, large areas of forest were cleared for oil-palm plantations. At this time too, systems of production that involved cruder processes such as the extraction of natural resources from forests were also prominent, for example rubber collection (particularly in the Congo Free State, formerly the Belgian Congo). Gold and ivory were particularly significant for the purposes of trade, even before the slave era.

For a short period at the turn of the nineteenth century, during the height of the Belgian King Leopold's hold on the Congo Free State, rubber extraction from the forest experienced a boom. People were forced to migrate to the forest, but the privations and pressures placed upon them to gather more and more rubber were so great that the resource was soon over-exploited. In the space of six years (1900 to 1906) the rubber resources of the 125 thousand km^2

Abir concession in the Congo Free State were completely exhausted (Reader 1997). Nevertheless, despite these efforts to force laborers into the forest, such was the pressure on the Congolese people that the population of the Congo actually decreased from an estimated 20 million, when the majority of concessions were granted, to only 8.5 million by 1911. In 1903, the British Consul to the Congo Free State noted that a "...region once populated by 40 thousand people now held only one thousand; in another, some 6 thousand people had been killed or mutilated in a period of six months" (Reader 1997: 579). These numbers need to be seen in their temporal context; in 1500, the total population of Sub-Saharan Africa has been estimated at only 47 million (Reader 1997), less than the current population of the Democratic Republic of the Congo (DRC). By the beginning of the twentieth century it had only risen to 129 million. While these are only estimates, they do suggest that depopulation was catastrophic, amounting to almost 9 percent of the total population of Sub-Saharan Africa.

Predations upon indigenous populations, extreme as they seem, were not atypical of the colonial period. Coquery-Vidrovitch (1985) for example, records mass movements of workers, coerced by the colonial state, throughout the French territories and in the central African colonies of the Belgian state. Indigenous people were forced to work on new cocoa and oil-palm plantations in the forest zone, savanna zone crops in the drier interior, and in mines. Large-scale movements of people accommodated these colonial demands, which continued through the Great Depression of the 1930s and into the period of the Second World War. Thus, these types of movements occurred relatively recently. Either because indigenous people succumbed to the recruitment drives of colonial companies or evaded them, many, perhaps tens or hundreds of thousands of peasants, fled or left their homelands. For example, large numbers of Rwandans were recruited for the Katanga mines, and in times of famine, multitudes emigrated. Thus, as Coquery-Vidrovitch (1985) states, 25,000 Rwandans left for Uganda during the 1928–29 famine. In West Africa, largely to avoid taxation, a similar number left the Mossi plateau of Upper Volta (Burkina Faso) for the Gold Coast (Ghana). In Gabon, vast forest concessions necessitated labor for logging. There are many other examples throughout West and Central Africa, all pointing to movements of large numbers of people throughout the zones and sectors of colonial exploitation. Many of these activities (plantation cropping, mining, logging) were located in or near the forest zone and it is safe to assume that many migrants stayed in these pioneer zones, albeit mostly in towns and around centers of employment, rather than settling on farms.

The extent to which these migrants inflated preexisting forest populations is unknown. Until quite recently, the humid forests of West and Central

Africa were sparsely peopled. Caldwell (1985) argues that 500 years ago, this vast forest zone was home to very few people, except on the fringes, along well-established trade routes, and along the coast. Rapid population increases in the nineteenth century were almost certainly due to natural increase rather than immigration.

When these colonial-instigated movements of people are added to the long disruption caused by the Atlantic slave trade it becomes clear that the humid forest zone of Africa has witnessed a constant movement of large numbers of its inhabitants for more than five centuries.

It is somewhat ironic that much of the current migration into and through the Amazon as described by Perz (Chapter 2) involves many people who can trace their origins back to the very same slave trade that initiated the great upheavals in Africa. The historical legacy of the slave trade reverberates into the present day. For example, Mogba and Freudenberger (1998) describe the current influx of people into the forest zone of southern Central African Republic (CAR) and locate it within the context of the long history of slave-raiding in the early nineteenth century. Despite the constraints that national boundaries place upon movement, particularly for agricultural settlement in new lands, people do move across frontiers and establish economic footholds (for example, the migrants from Burkina Faso who have settled in Côte d'Ivoire as cocoa growers). Such movements can often be due to non-agricultural activities, as in the forest of the CAR, where immigrants from throughout West Africa—even as far away as Senegal—have moved to mine diamonds.

Contemporary migration and demographic processes

Waves of migration in Africa vary in character and causation; they are not all driven by the same forces and do not have the same type of impact upon forestlands. In the past, there is little doubt that mass movements were engineered by colonial powers and their subsequent modes of exploitation. Laborers were coerced into supporting the extraction of minerals, timber, rubber, and other forest products. We do not know the figures for those who stayed on as the colonial economies shifted to crop production, but evidence from the Congo Basin suggests that many died and many must have desired nothing more than to return at the earliest opportunity to their homelands. A large number were thus unwilling migrants in search of new opportunities and new frontiers.

During the early twentieth century, as the colonial economies matured and became more structured, agricultural production of cocoa, cotton, and other products required by the metropolis began to replace the more overtly extractive systems that had predominated at the turn of the century and during the early 1900s. Peasant farmers began to move towards these centers of agricultural production initially due to taxation reasons but were later motivated by perceived opportunities for personal benefits and thus moved voluntarily. Painter (1985) suggests that the 1944 estimate of 200,000 French subjects (largely from Burkina Faso and western Niger) living in the then Gold Coast was probably an underestimate. Approximately half of these people remained to become permanent residents, largely in the forest zone. A similar evolution occurred in the Côte d'Ivoire. Indeed, much of the current unrest there is due to the uncertain status of these migrants who have become permanent settlers with some land tenure security but with apparently uncertain citizenship rights.

The pattern and logic of these movements changed dramatically as the colonial era came to an end and the new independent states established their territorial hegemony. Border control, along with the desire to build national identities, gradually narrowed the window of opportunity for migration. While movement within individual states continued, cross-border migration became progressively more difficult.

In the following section I discuss some of the major factors that affect migration trends in Sub-Saharan Africa. There are factors not directly linked to any forest resource. These include processes of reoccupation of indigenous lands—often spontaneous or forced relocations that resulted from abandonment of colonial territorial policies after independence. Two additional factors are processes of urbanization and demographic transition. A fourth and most important factor is the mass movement of refugees fleeing conflict situations. Migrations linked to these factors have had different degrees of impact upon forests, as discussed below.

Restitution of lands

The large-scale waves of transmigration that took place in Southeast Asia (see Potter, Chapter 8) never occurred as such in Africa. The closest equivalent is the restitution of lands following independence in the former settler economies of east and southern Africa. Most have now been completed, perhaps with the exception of Zimbabwe. However, on this smaller scale, the colonial settlement and the postcolonial readjustment that followed has displaced and then relocated considerable numbers of people. For Kenya, as for several other settler colonies, the independence struggle was all about land. Following Kenya's independence in 1963, the new government disbanded Native Reserves and implemented a program of land redistribution. Kenyan farmers moved onto lands abandoned by departing Europeans. Between 1969 and 1979, almost half a million people relocated. To a lesser extent, Tanzania followed the same trajectory of relocation. Zimbabwe is a late-comer to this pattern and is still in the throes of major disruption.

Very little of this forced migration took place in or near the evergreen forest zone; rather, it encompassed broad swathes of dry deciduous forest (*miombo*) and savanna woodland and, in the case of Kenya, some areas of montane forest. Tens of thousands of families were resettled on formerly white-owned estates in Kenya during the 1960s, but this was largely on land already occupied and in most cases, particularly recently, was rarely covered with dense forest. A little later, in Tanzania, the Ujamaa program reordered settlements into villages, a type of "villagization" that has also been seen in Zimbabwe and other parts of southern Africa. In some instances this relocation of local farmers—for the purposes of "betterment"—has led to the widespread clearance of trees, but this has occurred primarily in savanna woodlands rather than forest. Under the guise of better resource management, people living in small hamlets and scattered small cultivation plots along the valleys were forced to abandon their homes. Families were regrouped and resettled on the interfluves. New farmlands were carved out of the bush, leading to considerable clearance of *miombo* woodlands (Scoones 1990).

Urbanization and demographic transition

In Africa, ever since the modern urban pattern was developed during the colonial period, Africans—like peoples elsewhere in the world—have moved to towns in search of opportunities. In the case of the former French territories, the first surge in urban population growth occurred in the early 1930s when Dakar and Abidjan grew by 70 percent in the space of only five years, and this at a time when rural populations hardly grew at all (Coquery-Vidrovitch 1985). Similar growth rates accompanied the rise of industrial and mine-based urban centers throughout the continent. Nevertheless, despite this early start to urbanization, the rapid growth rates that characterized population dynamics in the middle colonial years ensured that rural populations also grew quickly. The expansion of towns did not lessen the growing pressure on rural land.

In the last thirty years, urbanization rates have accelerated further and at the current time show no signs of abating. According to the United Nations Economic Commission for Africa (UNECA 2003), over one third of Africa's people lived in towns in 1995. By 2015, it is anticipated that this proportion will have increased to 50 percent or more. However, individual countries show markedly different rates (Figure 5.2) and at the present time, total

Figure 5.2. Urbanization in Africa's humid forest states, 1965–1995

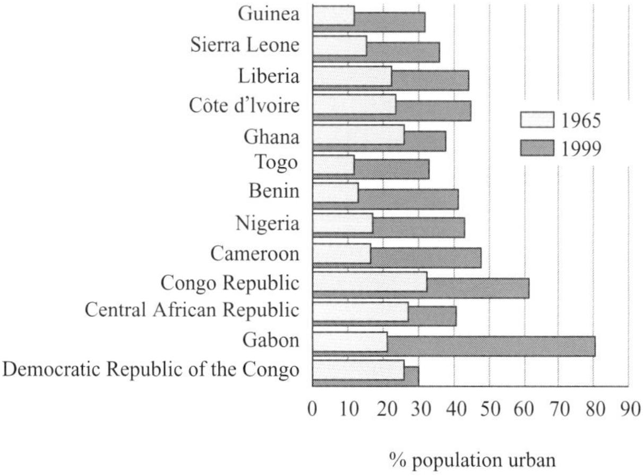

Source: World Bank http://www.worldbank.org/data/wdi2002/pdfs/table%203-10.pdf.

population growth subsumes this surge in urbanization. The movement of rural people to cities and towns as permanent or seasonal migrants has had little impact on rural populations thus far because rural population growth rates, often in excess of 3 percent per year, more than offset the exodus to cities. Figure 5.2 shows recent levels of urbanization for the major forest states of West and Central Africa. In all cases there has been a substantial increase in urbanization since 1965, particularly so in West Africa. In the Congo Basin, some states, such as Gabon and the Congo Republic, follow or even exceed the trend in West Africa, but for the DRC and the CAR urbanization rates are still comparatively low. The picture is compounded by the refugee problem that besets this sub-region. In fact, these data are not based on "solid" empirical work. In all likelihood, they are projections from previous surveys. Nevertheless, it is safe to accept that urbanization rates are climbing. The question for the future is the extent to which this flux of people into cities and towns will overtake rural population growth rates. Current projections suggest this is some way off, so that the pressure on land in rural zones will increase rather than diminish.

The foregoing considerations are mediated by changes in birth and death rates. There is growing evidence to suggest that birth rates are beginning to decline over much of Africa. Demographic transition models suggest that in a number of states populations will grow only slowly or stabilize. Figure 5.3 shows current projections. The chart is divided into four zones, corresponding to the major stages in the demographic transition model.

Figure 5.3. The demographic transition in Africa 1990–1995

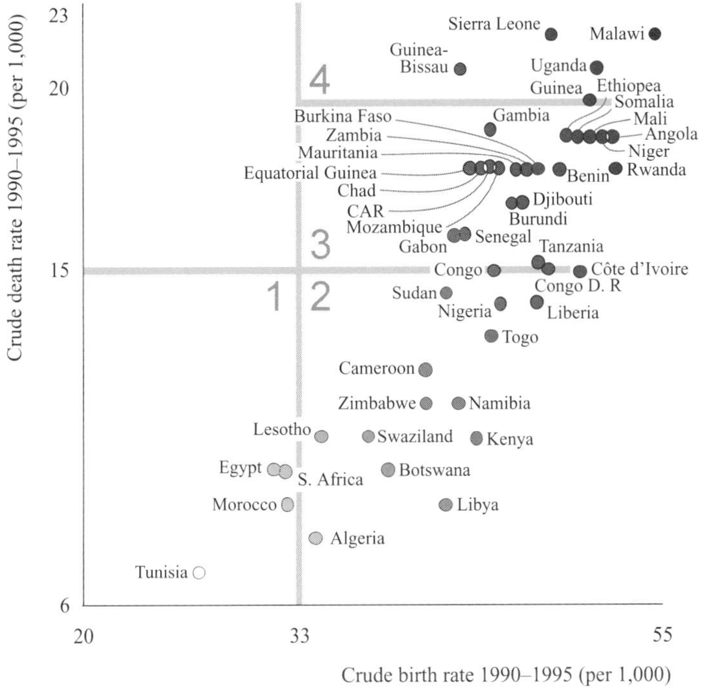

Source: World Bank data from Health, Nutrition and Population http://www1.worldbank.org/hnp/.

Thus zone 4 (high birth and death rates) and zone 3 (declining death rates, and yet high birth rates) includes many of the forest zone states. However, Nigeria, Cameroon, and Togo are located in zone 2 where birth rates are beginning to decline.

Refugees

A special pattern of migration springs from the recent escalation of civil unrest and insurrection across several regions of the continent. Very large numbers of refugees have moved in and out of forests as they flee from war zones. Though most head for the perceived safety and security of towns and cities, many flee to the forest where they can only exploit the limited resources available to them. Thus, as Mogba and Freudenberger (1998: 106) indicate in the context of the CAR: "tensions and social crises that accompany the processes of political change have been rife in several central African nations (including Rwanda, Burundi, and the DRC). The violence of military conflict has forced human communities to take refuge in the forests, where they have been living off the floral and faunal species there."

Figure 5.4 illustrates the patterns of movement from three centers of unrest: the DRC, Sierra Leone, and Liberia. These data are derived from the World Refugee Survey 2002 undertaken by the U.S. Committee for Refugees

Figure 5.4. Refugee movements out Liberia, Sierra Leone, and the Democratic Republic of the Congo

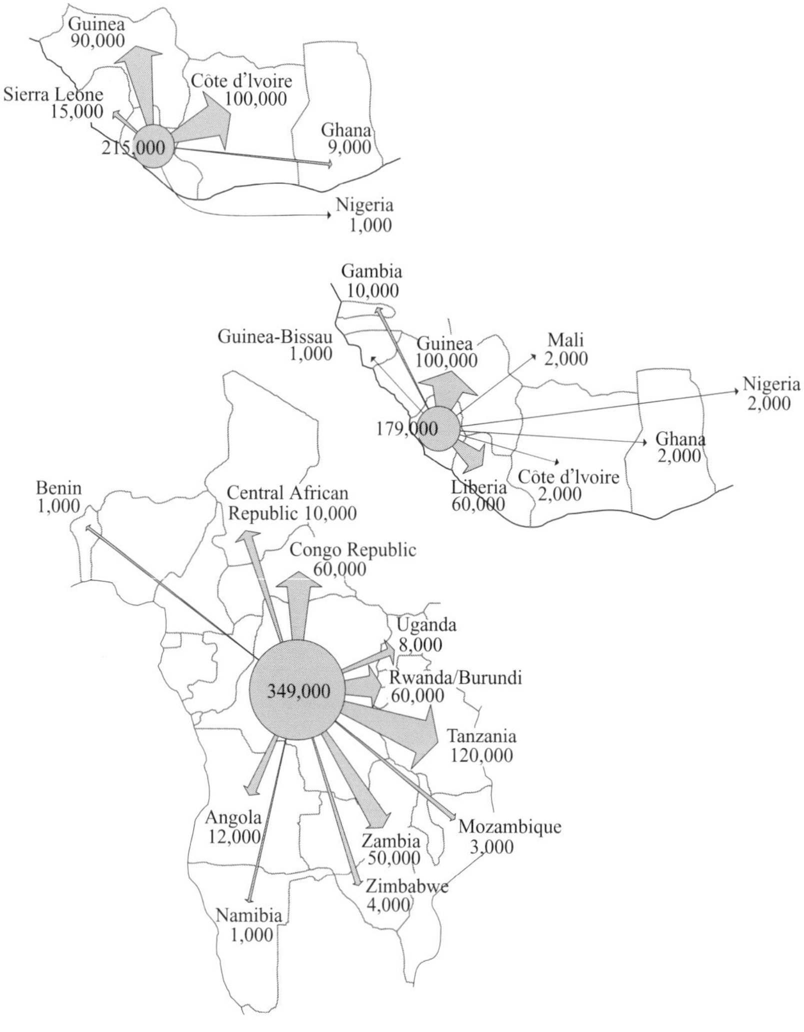

Source: World Refugee Survey 2002 (U.S. Committee for Refugees 2002).

(USCR 2002). Since then, violence has erupted in the Côte d'Ivoire and again in Liberia. As a result, yet more refugees can be expected to escape to Guinea and surrounding states.

It is hard to overestimate the sheer magnitude of Africa's refugee problem. Eleven African states have refugee populations of more than 100,000 (Crisp 2000). Six of these (the DRC, the Congo Republic, Guinea, Côte d'Ivoire, Uganda, Kenya) are situated in the humid/montane forest zone and a further two (Zambia, Tanzania) are neighboring countries with humid/montane forest zones. However, though the number of foreign refugees is large, it is sometimes dwarfed by the number of internally displaced people (IDPs) in a country. For instance, in the DRC there are an estimated 2 million IDPs (USCR 2002). Biswas and Tortajada-Quiroz (1996) note that, following the arrival of 1.5 to 2.0 million Rwandan refugees in Zaire, there was a loss of 3,758 ha of forest within three weeks in South Kivu (Zaire) and 300 ha in North Kivu, with a further 11,875 ha severely affected. Damage starts with the removal of dead wood, followed by the cutting of branches, then the whole stem and finally the roots. Soil erosion follows. The authors also note greatly increased poaching from the forest and the arrival of cattle into refugee zones. Apart from this brief study, little is known of the interaction between refugees and the forest. Severe impacts can only be imagined, however, as desperate people exploit whatever resources they can find.

A second sub-region beset by internal strife and displaced people is the region in West Africa incorporating Sierra Leone, Liberia, Côte d'Ivoire, and Guinea, again in the forest belt and in this instance forming the last extant remnants of the West African forest block. The flux of people among these countries is shown in Table 5.2.

The discussions in this section paint a complex contemporary picture. Changes in birth and urbanization rates, compounded by the refugee situation (including internal displacements), make the future uncertain. In the short-

Table 5.2. Movements of people within the Sierra Leone, Liberia, Côte d'Ivoire, Guinea complex

Location	December 1996	June 1998
Liberia	1,800,000	209,000
Sierra Leone	609,000	300,000
Côte d'Ivoire	305,000	140,000
Guinea	536,000	614,000
Total	3,250,000	1,263,000

Source: International Federation of Red Cross and Red Crescent Societies (cited 2003).

term, demand for new land will continue to put pressure on remaining forests. This issue is further complicated by the problems of trans-border migration and land tenure laws. What is certain is that Africa has experienced a long history of migration, in which the slave trade and the subsequent colonial period played their dramatic parts. Even with the solidification of modern state boundaries, pressures that have historically forced migration continue to be influential through refugee flows and the search for opportunity in cities and for new farmland. The search for farmland and economic profit from various forest products constitutes additional factors that cause migration patterns and profoundly affect the forest landscape in Sub-Saharan Africa, as the next section will show.

Migrants and resources in the forest realm

The rigid classification of forest resource use throughout history risks obscuring the complexity and variety of motivations that underlie migrations. Nevertheless, it is helpful to group these resource uses and the processes of movement associated with them, if only to appreciate their potential and differential impact upon the forest. There are several thematic dimensions to such a typology, including time, the numbers of people involved, the products they seek, and the extent to which these migrations are the product of internal (perhaps local) forces or of much broader external factors. While in all cases it is safe to assume that people move into and interact with the forest for individual and personal reasons, the wider pressures that place them in these situations must also be examined. The following thus attempts to encompass these dimensions.

As in so many locations, most migrants look for agricultural land in the forest. This has been and continues to be a primary factor affecting key forest areas, although the nature of this resource use and its related migratory processes is changing rapidly. Migrants attracted by agricultural lands can be grouped into three categories: permanent settlers, seasonal migrants, and so-called squatters. Permanent occupation is changing rapidly because of increased control and declining opportunities. Seasonal migrants may migrate across borders or inside national boundaries. Their contribution to the expansion of the agricultural frontier is, however, considerable (Acheampong, Chapter 6). A group of migrants related to both permanent agricultural settlement and seasonal labor are the squatters, in most cases "illegal occupants" of state owned forestland. A key factor that attracts migrants to the forest is the presence of resources with an international demand: timber, minerals, and bush-meat.

Permanent agricultural settlement

One resource use of key consequence for forests is agricultural production. The primary actors in this process in Africa are individual households or small groups seeking new lands to till as their homelands become crowded and arable land becomes scarce. This pattern has been the quintessential history of African migrations through which cultivators gradually populated the entire continent over several millennia. Cropland expansion was rapid; in the Côte d'Ivoire alone 5 million ha of forests were converted to croplands between 1966 and 1980 (World Rainforest Movement 2001: 3). This expansionism continues today, although in an arrested and restricted fashion. It is hindered by the territorial integrity of nation-states, a newly emergent continent-wide land shortage, and the restrictions arising from modern land laws. It is no longer so easy to cross frontiers and claim land-rights. Movements continue within individual states, as Acheampong (Chapter 6) describes in his account of the cocoa zone in Ghana, but even these internal movements are becoming more and more difficult. There is virtually no "spare" land left anywhere and under the present conditions of shortage—experienced not just by the would-be migrants themselves, but also by indigenous farmers—new settlers are generally not welcome.

As rural populations increase, the demand for farming land expands proportionately. Although urbanization rates are rising and some form of demographic transition is emerging, it is unlikely that in the near future these long-term processes will reduce the pressure on land resources. The last fifty years have seen the gradual degradation of the West African forests, but this is not simply the result of migrants clearing forest for new agricultural production. As Fairhead and Leach (1998) have shown, new forest is often "created" in this process, as villagers plant trees to protect the village environment from fire. Furthermore, migrants may come from afar (as in the case of Burkinabe settlers in the Côte d'Ivoire), or may be the result of gradual local expansion. It is increasingly difficult for people to move across frontiers for this purpose as modern land law constrains tenure security. Arguably, this continuing agricultural expansion is being driven by push-factors.

Seasonal labor

Somewhat related to agricultural settlement is seasonal labor. We can distinguish several types of movements within this complex. What Raynaut calls the "poles of attraction" (1997: 69) continue to draw seasonal labor from the Sahelian states. The present patterns of labor migration particularly from Burkina Faso to the cocoa and forest zones of the Côte d'Ivoire and

Ghana have continued since the 1920s. Migrants move to the forest zone and work on established cocoa farms and presumably also in a range of activities associated with both urban and rural economic sectors. They are attracted because of the very limited dry-season opportunities in their homelands. In 1985, 700,000 migrants left Burkina Faso primarily for seasonal employment, and many more must have left from Mali and western Niger (Sirven 1987). These types of migrants, sometimes seasonal and temporary and sometimes permanent, also traverse the dry zones, principally to the groundnut basins of Senegal and Nigeria, but also to the cotton belt in Mali. However, none of the dry zone migrations affect the forest zone.

Squatters

A relatively new phenomenon in forest-related migration is what foresters like to call "squatting". Obua and Aguma (2002) report on this at length for Uganda, where state forest reservations attract migrants and the landless as existing farmlands become fully occupied. Where no "spare" land exists for settlement, struggling families increasingly target state forestlands. This pattern of "invasion" is evident throughout eastern and southern Africa; in Kenya, Uganda, Zimbabwe, Zambia, and elsewhere. This phenomenon is exacerbated by the legacy of colonialism and the history of independence wars. Most of these struggles were concerned with land expropriated in the settler colonialism of the early twentieth century. Independence struggles, centered on reclaiming stolen lands, fueled the aspirations of those fighting for liberation. Within this consciousness, forest reserves were seen as part of the pattern of colonial land expropriation, no less than white-owned farms or estates: if the latter could be reclaimed, why not the former? Forest policy in the region has only recently accepted the principal of shared resources, however, during the first years of independence preexisting colonial policies of exclusion were carried over into the new era. The political costs of evicting squatters from forest reserves in Zimbabwe have been exposed in the modern politics of participation, where aspiring politicians are only too ready to target forest departments as enemies of the people (Bradley and McNamara 1993).

Extractive modes of production

In West Africa, Europeans never settled the land as farmers. Similarly, Central Africa and the Congo Basin were sparsely settled by European farmers and as a result none of the large-scale displacement of indigenous inhabitants that results from land seizure on behalf of settlers occurred. The humid forest has always yielded products and services for both domestic and international markets. Gold and ivory from the West African forest has been traded for over a thousand years, powering the rise of Sudanic empires from the tenth century onwards and more recently the mercantile sea-born trade of the pre-colonial period. Later, investors extracted other minerals; most recently and of greatest value, tantalum. In the colonial era, rubber and other non-timber forest products (NTFPs) were harvested, frequently to the point of exhaustion. This involved the forced movement of large numbers of people under great duress. Timber extraction is particularly damaging to the forest fabric, but the degradation of forest ecosystems by the burgeoning bush-meat trade that follows in its wake is also particularly conspicuous.

At the present time, transnational migrants continue to be attracted to these areas, mostly by non-farming opportunities. In some cases people migrate because they are pushed from their homeland. These push-factors include events such as the drying out of the Sahel and increasing landlessness. For example, in the southern CAR, set against a background of timber extraction which has continued unabated since the late 1940s, roads, small settlements, and access nodes have become established, leading to the immigration of considerable numbers of people seeking economic opportunities (Mogba and Freudenberger 1998).

Forest is commonly "opened up" by external agents (logging companies, mining interests, road or other infrastructural developments often funded by overseas aid) as the result of the "pull" of new opportunities. As access and opportunity develop, people move in, often from a great distance. The instance of small-scale mining intrusion into the southern forest of the CAR, following the routes opened up by timber concessions (Mogba and Freudenberger 1998) is a good example of this; it characterizes a mode of extraction that has continued for centuries and carries over, almost certainly in a more extensive pattern, into the modern era. This is a "frontier" movement, a continuous process that may be only temporary in nature and impact or that may develop into permanent settlements and irreversible clearance.

In addition to formal large companies, informal diamond extraction has expanded into many forest regions. The small-scale, informal sector develops mining camps that exploit the alluvial deposits to exhaustion, at which point camps are abandoned. These camps can be quite large; for example, Camp Zaire, mentioned by Mogba and Freudenberger (1998), grew to 2,000 residents before being deserted. As new camps are opened deeper in the forest, old ones are left behind resulting in a string of abandoned clearings.

Mogba and Freudenberger (1998) describe the influx of people into the Dzanga-Sangha Forest Reserve in the CAR. There (and doubtless elsewhere in the West and Central African forest zones), diamond mining combines with timber extraction to reveal a long history of penetration and exploitation. Following the sale of a diamond mining concession, the opening up of the mining zone attracted large numbers of unemployed rural and urban youth. The new arrivals mixed with indigenous forest peoples to exploit forest-based resources for both subsistence and commercial purposes. Small settlements developed and intermediary industries arrived to provide credit and a range of services—shops, alcohol, and machine repair. People required food and in response herders from the north brought livestock for sale. In the process they burned the forest edge to rid the area of tsetse flies, further opening up the canopy. Forest resources such as raffia and bamboo were exploited for building and the construction of tools. Soon agriculture developed; bushfires cleared emergent vegetation around the camps and villages and what was once forest became fields of maize and cassava. Bush-meat was harvested (poached) and animal resources quickly depleted.

Whilst little forest remains outside protected reserves in West Africa, extensive high quality forest remains unprotected in Central Africa. A scattering of reserves exist and there is considerable international pressure to conserve primates and other forest-based mammals. Nevertheless, despite official status, most of these reserves are poorly protected. Governance is weak and government agents corruptible. Extensive logging, mining, and bush-meat poaching continue in these areas, particularly where civil strife has lead to an influx of desperate refugees, as discussed above.

Bush-meat trade is now a multi-million dollar international business, with estimates of the total value of the African trade put as high as US$2 billion (Rose 2003). Ninety tons of bush-meat enters Yaoundé's markets each month (*Economist* 2003). In Gabon, the unregulated bush-meat trade is worth US$50 million per year (Steel 1994), although Inamdar *et al.* (1999) state a lower

Photo 5.1. Following tsetse fly eradication in West-central Zimbabwe in the 1980s, informal migration into the *miombo* has led to widespread forest clearance

figure of US$25 million per year. In the Congo Basin annual consumption is thought to exceed 1 million tons (Rose 2003), a figure supported by Inamdar *et al.* (1999) who estimate 1.2 million tons per year for the whole of the Central African forest region; the DRC accounts for almost 90 percent of bush-meat consumption in this region.

Bush-meat trade is directed to both urban domestic and international markets and is closely linked to logging activities. As well as moving timber to the ports, truckers transport great quantities of bush-meat supplied by local hunters from the logging concession areas. The bush-meat trade is also linked to other extractive industries such as mining, especially illegal tantalum mining deep in the Congo forest. The pre-eminent role of the DRC in the bush-meat trade is discussed at length by Yamagiwa (2003).

Forest cover and change

Statistics can be afforded greater authority than they deserve, potentially promoting inaccuracy and imprecision. There are many examples of the widespread acceptance and use of suspect "standard statistics" in the fields of forestry and deforestation (Fairhead and Leach 1998; Cline-Cole 2000). Continuing arguments about what "forest" (or, better, "forest cover") actually is reveal the false sense of authenticity that statistics bring to forest debates.

Forests in Africa include dry zone forests, such as the remnant teak forests of western Zimbabwe and Eastern Botswana which survive where annual rainfall is less than 1,000 mm. Some of the dense *miombo* woodlands of Zambia—in slightly more humid climates—are sometimes considered forests, but are also found in savanna climate zones. Areas with annual rainfall exceeding 1,500mm contain the rainforests of the West African coastal zone and the Congo Basin. Whether "forest" is defined as dense woodland or as true forest, cover estimates can be problematic. Further problems arise where temporal data is used to indicate the advance or retreat of forest cover.

The Global Forest Resources Assessment 2001 (FAO 2003) bases its estimates on more than 650 forest definitions collected from 132 developing countries. Many of these definitions do little more than note "closed forest", "open forest", or "other wooded land types". The FAO threshold of 10 percent canopy cover would incorporate the vast swathes of more open wooded savanna that characterize much of Africa's sub-humid regions (see Figure 5.5). The United Nations Environment Program-World Conservation

Monitoring Centre (UNEP-WCMC) adopt a more demanding 40 percent canopy cover to define forests (UNEP-WCMC 2001), whilst the World Resources Institute's Global Forest Watch (GFW) initiative and Central African Regional Program for the Environment (CARPE), using data from the Tropical Ecosystem Environment Observations by Satellites (TREES) studies, prefer a 68 percent canopy cover threshold to determine forest in Central Africa.

Many early surveys of forestland in Africa were in fact based solely upon surrogate measures (land areas lying within specific climatic limits, agricultural land-use data). A poignant example comes from Ghana, where Fairhead and Leach (1998) show estimates ranging from 5.44 to 9.90 million ha. Bradley and McNamara (1993) show similarly widely varying data from Zimbabwe. In theory, forest cover estimates ought to be more reliable when based on remote sensing imagery. CARPE, for example, now defers to satellite imagery wherever possible. Fairhead and Leach (1998) have argued, however, that even where high resolution data can be used, small scattered fields may be difficult to identify. Ground confirmation is at best patchy and may fail to resolve problems of image interpretation.

Estimates of present forest cover suffer from all the methodological problems outlined above. Nevertheless, modern surveys based on improved satellite imagery provide a workable approximation. The principal sources are the FAO, UNEP-WCMC, WRI-GFW, the International Union for the Conservation of Nature (IUCN), and the various agencies and programs working under the CARPE and TREES umbrellas. Table 5.3 shows a selection of comparable data for 1990.

The total for the two humid forest sub-regions is thus somewhere between 17 and 20 million ha. FAO estimates range from 19 to 24 million ha, depending on the survey date and methodology employed. More detailed data is also available for specific areas. Thus for the Cameroon,[3] GFW (2000b), calculates forest cover as 22,794,000 ha (including degraded forest). This report, along with similar data from Laporte *et al.* (1997) and Mayaux *et al.* (1998) bases its findings on the TREES project.[4] A series of maps have been constructed from these and other data. Together they provide the best and most accurate picture of tropical Africa's forest cover. Figure 5.5 shows forest cover across a range of canopy cover classes for the whole of West and Central Africa.

It is clear from both the tabled data and the map that the West African forest is severely depleted when compared to the contiguous block of the Congo Basin. In West Africa, we see little more than remnants; only forests protected in reserves remain in good condition. In the Congo Basin, on the

Table 5.3. Forest cover estimates for West and Central Africa

Country	Evergreen and semi-deciduous forest	Closed broadleaved forest	Closed forest
Source	TREES	FAO	IUCN
Equatorial Guinea	1,811	1,826	1,700
Central African Republic	6,037	7,824	5,224
Cameroon	17,378	7,428	15,533
Gabon	20,677	18,162	22,932
Congo	23,916	19,527	21,340
Dem. Rep. of Congo	114,147	103,533	119,074
Central Africa	183,967	158,300	185,802
Ghana	2,176	1,596	1,584
Guinea	1,754	1,626	469
Côte d'Ivoire	3,386	1,123	2,747
Liberia	3,524	4,633	4,124
Nigeria	5,635	5,613	3,862
Sierra Leone	963	680	506
Togo	266	248	136
Benin	156	50	42
West Africa	17,859	15,569	13,470

Sources: TREES: Mayaux (1998); FAO (1993, 1996); IUCN: Sayer *et al.* (1992).

Figure 5.5. Forest-cover classes for West and Central Africa

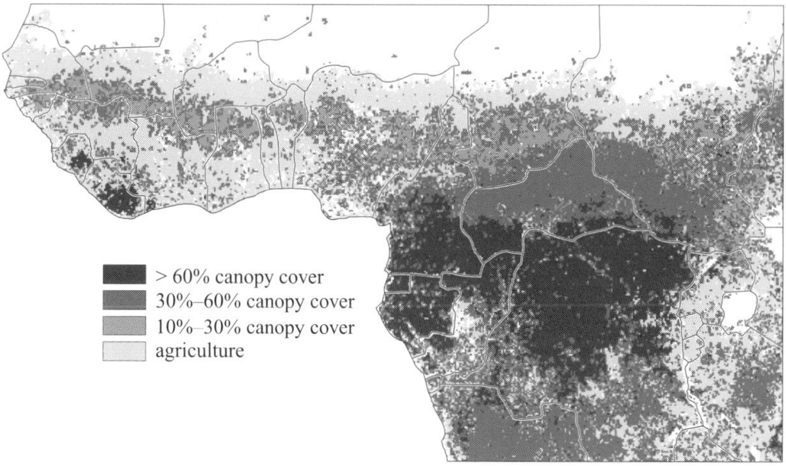

Source: Adapted from DeFries *et al.* (2000).

Table 5.4. Forest and population data for the West African and Central African forest blocks

	Forest cover (1,000 ha)		Annual de-forestation (1990–1995)	Forest cover (%)	Rural population (million)	Rural population density (person/km^2)
	2000	1995				
Central Africa: Congo basin	232,228	204,677	0.56	48.3	42.22	10.67
West Africa: humid forest zone	41,594	46,324	1.02	22.8	111.53	54.79

Source: FAO (1997, 2001).

other hand, extensive tracts of thinly populated and sparsely disturbed forest still exist. In West Africa the differentiation of forest from non-forest is unclear, including intergradations of secondary forest of varying stages and density, tree-clad farm plots, and recovering fallow areas. These gradations bear witness to the long and progressive clearance of forest for agriculture that has unfolded during the last two centuries in a region that has always been more densely populated than the Congo Basin block. Rural population density in West Africa is five times that of the Congo sub-region (Table 5.4). This, in part, explains the difference in forest cover.

The history of West Africa and the Congo sub-region is markedly different, the former having experienced a much more intense and prolonged exposure to pre-colonial and colonial contact with the rest of the world and a much greater involvement in foreign trade and a higher level of commercial intrusion than the latter. This too must have left its imprint upon West Africa's forests and is intimately connected to the demographic pressure experienced by the region. In the Congo Basin forests, population density is low. With the exception of the eastern rim, set against the rift valley volcanics and their associated fertile soils where population densities are high and intensive agriculture the norm, much of the northern Congo Basin forests are relatively empty and for the most part resist immigration for agricultural settlement. Soils are of poor quality and the general ecology is inhospitable.

Forest-cover change

Because the decline of forests is derived from area estimates of cover over time, the methodological problems that constrain our knowledge of current forest extent also hamper our understanding of change. This is significant, if only because forest destruction rates are too often linked to population growth and migration. There is insufficient space here to detail the difficulties of determining forest change (but see GFW 2000a, 2000b; Matthews 2001; TREES 2003). The principal problems can be briefly listed here:

- Climatic data are used to determine "potential" forest limits. This is particularly the case with the "derived" savanna of the West African forest block. This is generally regarded as a non-climatic formation, an "artificial" savanna, "derived" from its original forest by human action. It emerges that the limits of this zone were suggested by Aubreville (1938) who argued that as forest is cleared, forest climates become drier so that modern day climatic limits underestimate the true extent of the original forest. He calculated the old forest extent by inflating present rainfall figures by 15–20 percent. Much of the variation in forest-cover estimates for Ghana is explained by this construct.
- Forest-cover estimates for the distant past (c. 1,000 years or more) are little more than guesses. Most doubt centers on estimates of the original forest cover (that is, prior to the effects of modern man). For example, the WRI-GFW, in describing forest cover in Cameroon, speaks of an historic forest cover of 37,400,000 ha deemed to have existed 8,000 years ago, prior to large-scale human disturbance: "Over 18 million hectares of Cameroon's historic forests have been cleared to make way for agriculture and settlements" (GFW 2000b: 11). From these data, by 1995, the original forest cover had declined by 31 percent (Cameroon) and 48 percent (Gabon).
- Estimates of the recent past (c. 100 years) are often based on surrogate measures or very partial forest surveys.
- Interpretations of forest-cover change (both retreat and advance) are too often ensnared in "narratives" which have little scientific validity and have been eloquently exposed by Fairhead and Leach (1996), who warn of the dangers of assuming former forest extent based on relic elements (islands of oil-palm trees, silk cotton trees, etc.). They argue convincingly that in the Kissidougou region of Guinea at least, such islands represent expansion rather than retraction. A similar picture emerges from the Ivorian savanna where forest blocks are found to be relatively recent in origin (Fairhead and Leach 1998).

With these caveats in mind, the following data on the extent and deforestation rates of the two major humid forest blocks in Africa (West Africa and Central Africa-Congo Basin) should be treated with caution. Using 1 km^2 imagery at a broad scale and 20 to 30 m imagery on a sampled basis Achard *et al.* (2002) calculate that between 1990 and 1997, Africa's humid forests were retreating by

790,000 ha per annum—due to agriculture, to mosaics, and to savanna woodland. This translates to an annual depletion rate of 0.36 percent. FAO estimates are almost three times this rate. Between 1995 and 2000, some 22,821,000 ha were lost (FAO 2001). For the West African forest block the annual deforestation rate is calculated at 1.02 percent, whilst for the Central African-Congo Basin forest the decline is less rapid at 0.56 percent. The difference in dates is unlikely to account for the discrepancy between the two sets of estimates, which is almost certainly the result of dissimilar methodologies.

Figure 5.6 shows the location of "hotspots" of pressure in the two forest blocks. For Central Africa, pressure is most keenly felt near to the forest–savanna boundary along the northern margins, the eastern rift valley region, and in the southeast. It is certain that this pressure will be exacerbated by refugee flows and the general lack of civic order in the DRC. The relative lack of hotspots in the West African zone simply reaffirms the fragmentary nature of the remaining forest area. TREES's (2003) identification of "hotspots" draws attention to the process of forest intrusion and its outcomes, thus:

> The common pattern of forest colonization is always the same in the region. Firstly, logging companies, exploiting government concessions, open the forest cover with a reticulated network of primary and secondary roads. When there is a demand for food products (bush-meat, manioc, bananas) in the cities, people install new settlements along the roads and initiate the trade of such products, which generates colonization within the primary forest. However, if the population pressure is not high enough or if the centres of demand are too far away, forest tracks can close up in five years (Defourny and de Wasseige, cited in TREES 2003).

Figure 5.6. Hotspots in the West and Central African forests

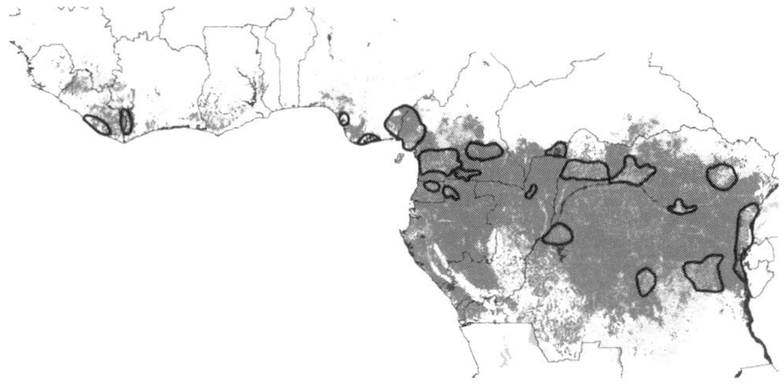

Source: Adapted from Achard *et al.* (2002).

In comparing the two forest blocks, it is too simplistic, however, to associate the greater population pressure in West Africa and its current poor forest cover with the result of a simple neo-Malthusian migratory push. The dynamics of forest clearance are complex and variable. Often, the initial opening of the forest stems from logging activity, as we have observed. Roads are created, minor settlement and service facilities follow, and migrants move in search of opportunities. This process of incursion is well demonstrated in the Dzanga-Sangha Forest Reserve in the southern CAR, mentioned earlier.

In the case of West Africa, there is a mix of pull- and push-factors. Whilst the new economic opportunities offered by the cocoa and coffee trade encouraged migration from the interior, each drought in the drier savanna and Sahelian zones produced a brief period of migration as conditions deteriorated in the north. Patterns of long-distance labor migration developed and these both underpinned and refueled the movements of people to the coastal forest zones. The current unrest in the Côte d'Ivoire, which is set to unleash another refugee crisis, is in part an outcome of these migratory flows.

Conclusions

Africa is both large and heterogeneous. Any attempt to review the social ecology of migrations and forests at a continental scale risks broad generalization which may make little sense at a more local level. However, a range of generic situations and processes have been described, enriched with local detail where possible. An important dimension to this review has been a focus on history and the impact of past events on current dynamics. It is also the case that much of our knowledge is based on data that is not as precise as it may at first appear. This is particularly true of data on the extent of forest cover and the rate and nature of deforestation, where recent studies have cast considerable doubt on what had previously been accepted as truth. For the future, there are positive signs that co-management of forest resources, in which local people become stakeholders of a valuable resource, may go some way to securing a long-term future for forest resources, although this is only ever likely to affect small areas. At the same time, the impact of the global economy, through mining and timber interests, through the expanding bush-meat trade (whose spatial dimensions are truly global, if informal), through the continuing impoverishment of the peasantry, which responds by continuing to seek out new land, does not bode well for a sustainable future.

Acknowledgements

The author would like to acknowledge the financial and logistical support of the Japan Centre for Area Studies National Museum of Ethnology, the Forestry and Forest Products Research Institute, and the Centre for International Forestry Research, along with the Center for South East Asian Studies, Kyoto University, Research Institute for Humanity and Nature, and the National Institute for Environmental Studies. All, in their different ways, have contributed to a stimulating and successful symposium. I would particularly like to thank the organizers and support staff of the symposium, especially Wil de Jong, Lye Tuck-Po and Abe Ken-ichi. Finally I thank Emmanuel Acheampong, Robert Aguma, Yamagiwa Juichi and Oikawa Yosei for waving the flag for Africa.

Notes

1. Area data from the Food and Agriculture Organization (FAO), Production Yearbook and data files. Population data from World Bank staff estimates from various sources including the United Nations Statistics Division's Population and Vital Statistics Report, country statistical offices, and Demographic and Health Surveys from national sources and Macro International.
2. The colonial administration recognized that plantations of this type were labor-intensive and would result in the coercion of thousands of peasants. They refused permission (Rodney 1985).
3. Equivalent detail exists for the other Central African forest states from the same sources.
4. Estimates provided by GFW, Laporte *et al.* (1997) and Mayaux *et al.* (1998) are all based on 1992–1993 NOAA-AVHRR data, with sampled areas verified with Landsat TM imagery.

References

Achard, F. *et al.* 2002. Determination of deforestation rates of the World's human tropical forests. *Science*, 297: 999–1002.

Aubreville, A. 1938. La fôret coloniale: Les fôrets de l'Afrique Occidentale Française. *Annales d'Académie des Sciences Coloniales, IX, Société d'Editions Géographiques, Maritimes et Coloniales*. Paris.

Biswas, A.K. and H.C. Tortajada-Quiroz. 1996. Environmental impacts of the Rwandan refugees on Zaire. *Ambio*, 25(6): 403–408.

Bradley, P.N. and K. McNamara. 1993. *Living with Trees: Policies for Forestry Management in Zimbabwe*. Washington DC: World Bank.

Caldwell, J.C. 1985. The social repercussions of colonial rule: demographic aspects. In A. Adu Boahen, ed., *General History of Africa. VII: Africa under Colonial Domination 1880–1935*, 458–487. London: Heinemann Educational Books.

Cline-Cole, R. 2000. Knowledge claims, landscape and the fuelwood-degradation nexus in Dryland Nigeria. In V. Broch-Due, and R.A. Schroeder, eds., *Producing Nature and Poverty in Africa*, 109–147. Stockholm: Nordiska Afrikainstitutet.

Coquery-Vidrovitch, C. 1985. The colonial history of the former French, Belgian and Portuguese zones, 1914–35. In A. Adu Boahen, ed., *General History of Africa. VII: Africa under Colonial Domination 1880–1935*, 351–381. London: Heinemann Educational Books.

Crisp, J. 2000. Africa's refugees: patterns, problems and policy challenges. *New Issues in Refugee Research Working Paper*, 28. Geneva: UNHCR.

DeFries, R., M. Hansen, J.R.G. Townshend, A.C. Janetos, and T.R. Loveland. 2000. A new global 1km data set of percent tree cover derived from remote sensing. *Global Change Biology*, 6: 247–254.

Economist. 2003. Trading in bush-meat—Africa's vanishing apes. *The Economist,* 10 January, 2002. http://www.economist.com/world/africa/

Fairhead, J. and M. Leach. 1996. *Misreading the African Landscape: Society and Ecology in a Forest-Savanna Mosaic*. Cambridge: Cambridge University Press.

Fairhead, J. and M. Leach. 1998. *Reframing Deforestation: Global Analysis and Local Realities: Studies from West Africa*. London: Routledge.

FAO (Food and Agriculture Organisation). 1993. *Forest Resources Assessment 1990: Tropical Countries*, FAO Forestry Paper 112.

FAO. 1996. *Forest Resources Assessment 1990. Survey of Tropical Forest Cover and Study of Change Processes*, FAO Forestry Paper 130.

FAO. 1997. *State of the World's Forests 1997*, Rome, FAO. Cited 2003: http://www.fao.org/docrep/w4345e/w4345e00.htm

FAO. 2001. *State of the World's Forests 2001*, Rome, FAO. http://www.fao.org/DOCREP/003/Y0900E/y0900e00.htm

FAO. 2003. *State of the World's Forests 2001*, Rome, FAO. http://www.fao.org/DOCREP/005/Y7581E/y7581e00.htm

GFW (Global Forest Watch). 2000a. *A First Look at Logging in Gabon*. World

Resources Institute, Washington DC. http://www.globalforestwatch. orgGFW. 2000b. *An Overview of Logging in Cameroon.* World Resources Institute, Washington DC. http://www.globalforestwatch.org

Inamdar, A., D. Brown, and S. Cobb. 1999. What's special about wildlife management in forests? Concepts and models of rights-based management, with recent evidence from West-Central Africa. *ODI: Natural Resource Perspectives* 44. London: Overseas Development Institute.

International Federation of Red Cross and Red Crescent Societies (IFRC). Cited 2003. *Refugees and Internally Displaces Persons (IDPs) in Africa.* http://www.ifrc.org/WHAT/health/archi/fact/frefugee.htm

Laporte, N.T., S.J. Goetz, C.O. Justice, and M. Heinicke. 1997. A new land cover map of central Africa derived from multi-resolution multi-temporal AVHRR data. *Int. J. Remote Sensing*, 19(18): 3527–3550.

Manning, P. 1988. The impact of slave exports on the population of the western coast of Africa 1700–1850. In S. Daget, ed., *De la traite à l'esclavage.* Actes du colloque international sur la traite des Noirs, Nantes 1985. Nantes: Centre de recherche sur l'histoire du monde atlantique.

Manning, P. 1990. *Slavery and African Life—Occidental, Oriental and African Slave Trades.* Cambridge: Cambridge University Press.

Matthews, E. 2001. Understanding the FRA 2000. *World Resources Institute Forest Briefing* 1. Washington DC: World Resources Institute. http://www.wri.org/wri/forests/fra2000.html

Mayaux, P., F. Achard, and J.P. Malingreau. 1998. Global tropical forest area measurements derived from coarse resolution maps at a global level: A comparison with other approaches. *Environmental Conservation*, 25(1): 37–52.

Mogba, Z. and M. Freudenberger. 1998. Human migration in the protected zones of Central Africa: The case of the Dzanga-Sangha Special Reserve. *World Wildlife Fund—US Bulletin*, 102.

Obua, J. and R. Aguma. 2002. Illegal settlements and evictions from forest reserves in Uganda: Implications for policy, governance and management. Paper presented at the symposium: *In, Out, In, Out: Populations, Migrations, and the Social Ecology of Tropical Forests.* Osaka, December 5–7, Japan Center for Area Studies.

Painter, T. 1985. *Peasant Migrations and Rural Transformation in Niger: A Study of Incorporation within a West African Capitalist Regional Economy.* Ann Arbor: University Microfilms International.

Raynaut, C. 1997. *Societies and Nature in the Sahel.* London: Routledge.

Reader, J. 1997. *A Biography of the Continent.* London: Penguin

Rodney, W. 1985. The colonial economy. In A. Adu Boahen, ed., *General*

History of Africa. VII: Africa under Colonial Domination 1880–1935, 332–350. London: Heinemann Educational Books.

Rose, A.L. Cited 2003. *Growing Commerce In Bushmeat Destroys Great Apes And Threatens Humanity.* http://bushmeat.net/afprimates98.htm

Sayer, J., C.S. Harcourt, and N.M. Collins. 1992. *Conservation Atlas of Tropical Forests: Africa.* London: Macmillan.

Scoones, I.C. 1990. *Livestock Populations and the Household Economy: A Case Study from Southern Zimbabwe.* PhD Thesis: University of London.

Sirven, P. 1987. Démographie et villes au Burkina Faso. *Cahiers d'Outre-mer*, 40(159): 265–283.

Steel, E.A. 1994. *Study of the Value and Volume of Bushmeat Commerce in Gabon.* WWF Programme pour le Gabon.

TREES (Tropical Ecosystem Environment Observations by Satellites). Cited 2003. *Deforestation Hot-Spots.* http://www.gvm.sai.jrc.it/Forest/Africa/hotspots.htm

UNECA (United Nations Economic Commission for Africa). Cited 2003. *Africa's Population and Development Bulletin June–July 2000.* Addis Ababa: UNECA. http://www.un.org/depts/eca/divis/index.htm

UNEP-WCMC (United Nations Environment Programme-World Conservation Monitoring Centre). 2001. *Global Distribution of Current Forest.* Cited 2003: http://www.unep-wcmc.org/forest/global_map.htm

U.S. Committee for Refugees (USCR). 2002. *World Refugee Survey, 2002.* Washington DC: US Committee for Refugees. Cited 2003: http://www.refugees.org/world/worldmain.htm

World Rainforest Movement (WRM). 2001. *Local Struggles and News Bulletin*, 43, February. Cited 2003: http://www.wrm.org.uy

Yamagiwa, J. 2003. Bush-meat poaching and the conservation crisis in Kahuzi-Biega National Park, Democratic Republic of Congo. *Journal of Sustainable Forestry*, 16: 115–135.

6
Migrations, Frontiers and the Forest Margin in South-western Ghana

Emmanuel Acheampong

Introduction

Local communities have played a major role in the expansion of agricultural production into the forests of Ghana, particularly during the period of the "great clearances" between the fifteenth and eighteenth centuries (Brown 1999). Historically, migrant farmers and laborers have been the primary agents of forest conversion and agricultural development. The influx of migrants into forest areas in search of empty land has caused massive deforestation. The creation of the country's forest reserves was partly a response to large-scale forest clearance for agricultural production, particularly cocoa. H.N. Thompson, conservator of forests in southern Nigeria, prepared a report for the Governor of the Gold Coast (now Ghana) in 1908 that led to the creation of a Forestry Department in 1909 with the aim of encouraging the reservation of 20–25 percent of the country's land area. The report identifies cocoa production and timber exploitation as the primary culprits of forest conversion (Smith 1999). Forest conversion and frontier farming continue to occur in the forest zone of southern Ghana, particularly in the Western Region.

In this paper I examine frontier farming and settlement in Ghana and their impact on forest resources. I begin with a brief discussion of the forests of Ghana providing a historical overview of frontier farming and forest settlement in the country. The discussion then focuses on an examination of the cocoa pioneer frontier of the Western Region. In subsequent sections I consider the impact of the influx of farmers on land tenure, land holdings, and agricultural production in the frontier, looking at how these factors shape local people's dependency on the remaining forest resources. In the final section I discuss the condition of the nation's forest resources and the extent of forest decline.

Frontier farming and forest settlement in Ghana

The vegetation of Ghana can broadly be divided into the dry savannah of the north and the high forest of the south. The high forest zone (HFZ) covers an area of approximately 9 million ha. However, it is estimated that about 7 million ha of this forest area have been severely modified over the last century and have no significant forest cover remaining (Brown 1999). Forest reserves represent the nation's permanent forest estate and produce the bulk of the nation's timber. The country has about 214 forest reserves covering approximately 1.8 million ha (Prah 1994). The Western Region features the greatest concentration of forest reserves, followed by the Ashanti Region and then the Brong-Ahafo Region[1] (Table 6.1, Figure 6.1). There are a few remaining areas of intact closed canopy forest outside reserves. Estimates of the area of non-reserve intact forest-cover vary from 3,740km^2 (Nsenkyire 1992) to 2,700km^2 (World Bank 1987) to 1,000km^2 (IUCN 1992).

Cocoa farming in Ghana has expanded into the HFZ through a process of pioneer frontier settlement—defined as "the movement of populations into new areas with empty, unsettled agricultural lands" (Amanor 1994: 33). Pioneer frontier settlement and forest farming in tropical forests is frequently associated with extensive shifting agriculture resulting in the rapid conversion of forest areas into agricultural fields (Richards 1985). A large percentage of small farmers in Ghana move from old zones of production to new frontiers to counter declining yields and land fragmentation. The historical pattern of frontier farming and the extent of migration into Ghana's HFZ has been extensively documented (Hill 1963; Johnson 1964; Wilks 1977; Robertson 1987; Amanor 1994). Frontier farming encroached into the forests of Ghana between the fifteenth and nineteenth centuries when large areas of sparsely

Table 6.1. Area of forest reserves by region

Region	Forest reserve area (square miles)	Total forest reserve cover (%)
Western Region	2,884	40
Ashanti Region	1,515	21
Brong-Ahafo Region	1,196	17
Eastern Region	677	10
Central Region	578	8
Volta Region	277	4
Total	7,130	100

Source: Ghana Forestry Department, cited in Brown (1999).

Figure 6.1. Regional map of Ghana

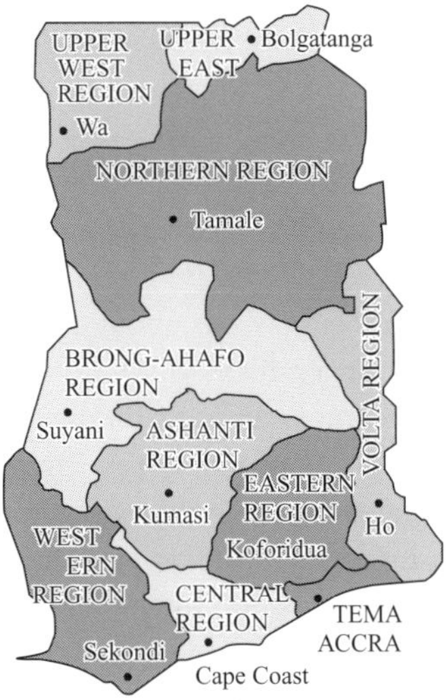

populated land were used for agricultural colonization (Wilks 1977; Amanor 1994). Immigrant labor was central to this expansion throughout this period. Initially, slave labor enabled the Akan, the largest ethnic group in Ghana, to clear the dense forests of southern Ghana for farming. The labor supplies were largely financed by the lucrative trade in gold bullion. Some farming on a very limited scale had probably been practiced in Ghana's forests for millennia. However, only when the Akan began importing slaves in the fifteenth and sixteenth centuries were they able to shift from an economy that relied primarily on hunting and gathering to one which became largely agricultural (Wilks 1977; Dounias and Leclerc, Chapter 7).

While this transition to agriculture took place, Akan communities continued to plant more of their traditional crops—plantains, yams, and rice—but they also adopted new crops from the Americas brought to Africa by Europeans, including maize and cassava (Wilks 1977). Farming led to a rapid increase in population in the forest region. As this population grew, small groups migrated across Ghana's forests in search of good farmland. Often entrepreneurs led these groups using slaves to do the initial work of clearing the forest. Later, these entrepreneurs would invite free settlers to join them, leading to new settlements throughout the forest (Wilks 1977).

During the early nineteenth century, pioneer frontier settlement was associated with the expansion of oil-palm production for export (Johnson 1964). However, two main factors caused the decline of the Gold Coast oil-palm industry: the decline of world palm oil prices during the 1860s and 1870s as a result of the discovery of mineral oil that replaced the industrial usage of palm oil, and the opening of European-owned oil-palm plantations in Malaysia and Java (Amanor 1994). This led to experimentation with new crops including coffee, rubber, and cocoa. Cocoa emerged as the most profitable crop in the Gold Coast becoming the focus of colonial economic policy (Amanor 1994). Cocoa production spread rapidly and by the end of the nineteenth century, this was the major contributor to frontier expansion. During the early twentieth century the Gold Coast emerged as the major world cocoa producer (Hill 1963).

Amanor (1994: 34) distinguishes five frontier zones in the forest area of Ghana: 1) the Eastern Region that developed in the early nineteenth century; 2) the middle frontier areas of the Ashanti and southern Brong-Ahafo Regions opened up from the 1920s; 3) the cocoa pioneer frontier of the Western and Central Regions opened up by farmers during the late 1960s; 4) the northern transition zone in the Brong-Ahafo Region that became the major food-producing zone during the 1970s; and 5) the oil-palm belt, situated within the cocoa belt of the moist semi-deciduous forest and wet evergreen forest within the Eastern, Central, Western, and Ashanti Regions that developed in the 1970s (Figure 6.1).

The cocoa pioneer frontier of the Western Region

The cocoa pioneer frontier of the Western Region is one of the newest frontier areas in Ghana. The history of cocoa is important in understanding the present social and economic conditions of production in this area, accounting for many features of tenure systems and labor relations. Historically, cocoa cultivation spread from the Eastern Region towards the Ashanti and Brong-Ahafo Regions in the 1940s and 1950s and further towards the frontier area of the Western Region during the 1960s (Quisumbing *et al.* 1999). From the Akwapim ridge where the crop was first introduced during the second half of the nineteenth century, the cocoa frontier first moved further north into Ashanti upon construction of railways and roads from the coast into the interior (Benneh 1988). By the end of the Second World War, the majority of the cocoa soils in the Eastern and Ashanti Regions had been cultivated. The near exhaustion of suitable cocoa land in these regions, combined with the devastation of cocoa by the swollen shoot disease in the Eastern Region, forced farmers to move into the less developed forest areas of the Brong-Ahafo, Central, and Western Regions (Benneh 1988).

The search for unclaimed land by pioneer cocoa farmers was also greatly facilitated by timber prospectors who built a network of bush tracks. The farmers' frontier, by and large, followed that of the timber prospectors in a region where there were hardly any roads (Benneh 1965). The expanding frontier absorbed a considerable proportion of the growing agricultural population. Thus, the Western Region is now the major cocoa growing area in Ghana and the principal area to which farmers and laborers are migrating. The area has the most heterogeneous rural population in Ghana, drawing the land-hungry from all other agricultural areas (Arhin 1985). However, as Arhin (1985) notes, the soils and ecology of this area are very fragile and the expansion of cultivation results in rapid degradation.

Smith (1994) reports that during the 1960s a number of forest reserves in the Western Region suffered massive encroachment. These include Manzan, Sukusuku, Bodi, Tano Ehuro and Bia-Tawya, all of which were converted to cocoa farms with well developed infrastructural and government services. By 1974 there were approximately sixty three different settlements, twenty-two government schools, and nine market centers in these reserves. Of the total forest reserve area of 1,637 km^2, some 999 km^2 contained illegal farms and 196 km^2 featured legitimate farms (Smith 1994).

Ethnic composition of the frontier

The study on which this paper is based was conducted in the Wassa Amenfi District in the Western Region of southwest Ghana (Figure 6.2). The area lies in the evergreen forest zone and falls within the jurisdiction of the Asankrangwa Forest District.[2] The total land area of the Wassa Amenfi District is approximately 4,747 km^2, which is about 17.8 percent of the total land area of the Western Region (WADA 1996). The district has extensive forest resources—part of the only surviving high forests of Ghana. Its eleven forest reserves cover 1,215 km^2. Data was collected in three forest communities in the district, namely Sureso, Kamaso, and Kamaboi, during a period of six months between January and June 2001 using both quantitative and qualitative methods. A total of 160 households were interviewed.

The ethnic composition of the study area as in many areas of Ghana's HFZ is highly diverse. Migrants are so widespread throughout the forest zone that indigenous people often do not comprise the majority of village populations. Although immigration has occurred more recently in the Western Region, migrant farmers often now dominate the local economy. In the three study communities, 27 percent of the 160 sampled households

Figure 6.2. Map of the study area in relation to the Western Region and Ghana

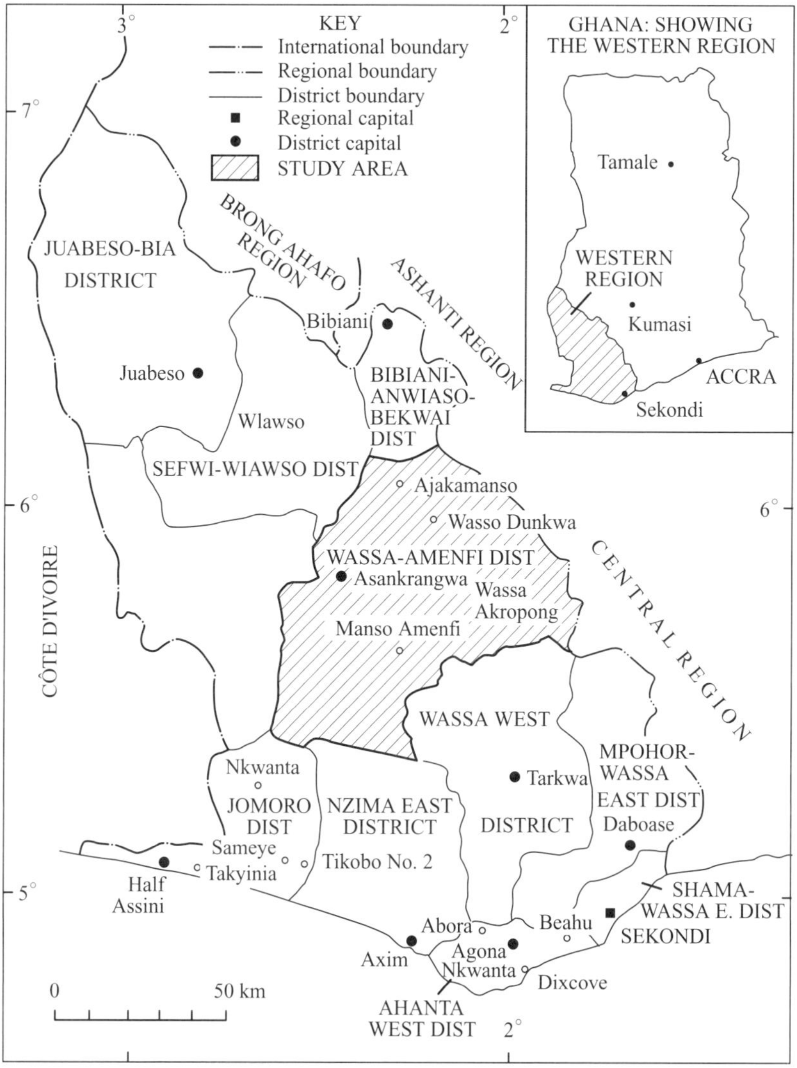

Source: WADA (1996).

were indigenes (the Wassas) while 73 percent were immigrants, including Asantes, Fantes, Nzemas, Kwahus, Akuapems, Akims, Ewes, Ga-Adangbes, and northerners from the Northern, Upper East, or Upper West regions. In the neighboring Sefwi-Wiaso District, with a population of 69,387, the latest official figures for registered migrants indicate 3,874 family heads, representing about 20,000 family members. Similarly, in the Juabeso District (also in the Western Region), with the total registered population of 129,101, it is estimated that migrants comprise approximately 45 percent of the resident population[3] (Brown 1999).

In these areas, three main categories of migrants have played and continue to play a central role in both the agricultural economy and in opening up the forest to further settlement: 1) southern migrant farmers who obtain land for agricultural production, often intending to plant cocoa; 2) migrant laborers, usually originating from the northern parts of the country, often taken on as sharecroppers; and 3) casual laborers and annual contract workers who are not fully settled but keep moving back and forth—primarily young men and women who travel to these areas to seek casual employment, mostly in cocoa farms.

The population of the entire Wassa Amenfi District has increased considerably with the influx of migrant cocoa farmers. In 1948 the population of the district was 21,500. By 1960 it had increased to 57,537 (an increase of 167.6 percent). In 1970 the population rose to 85,698. More than half (53.4 percent) of the population was born outside the district. The 1984 population census recorded 139,833 people for the district, an increase of 63.2 percent (Benneh 1988). The year 2000 population and housing census revealed a population of 229,421 persons for the district (GSS 2000), representing an increase of 64.1 percent over the 1984 figure.

The migration of farmers into this area to settle for the purpose of cocoa farming has placed severe pressure on land and forest resources, leading to complex changes in the agrarian and ecological systems. These issues are explored in the next section.

Land tenure issues at the frontier

The whole of the Wassa Amenfi District is under customary land tenure, where all land is ultimately controlled by village chiefs and family heads on behalf of the community, clans, and families. Lands under forest reserves, however, are vested in the government in trust for the landowning communities.

Individuals from the landowning communities often acquire land through family allocation, inheritance, and gifts. Gifts of land are usually made in appreciation of services rendered, and it is common for landlords to give portions of land to children or servants after they have worked for a substantial period of time on their farms. The mode of land transfer has

Table 6.2. Land ownership among migrants and indigenes (percent of respondents)

	Indigenes (%)	Migrants (%)	Total (%)
Farm on family land	39.5	20.5	25.6
Farm on individual land	53.5	51.3	51.9
Farm on hired land or sharecropping	7.0	28.2	22.5
Total number of respondents	43	117	160

evolved and changed over time. Quisumbing *et al.* (1999) note that when virgin forests were abundant, forests were appropriated primarily by young males before marriage mainly for the production of food crops. Relatively strong land-rights were granted in return for the substantial labor input required to clear forest. Traditionally, in the Akan inheritance system, this type of land was either bequeathed to nephews or allocated to other male members of the extended family in accordance with the decision of the family head. Wives and children were left with no rights to a man's property if he were to die (Quisumbing *et al.* 1999).

More recently, however, appropriated village land and even family land are increasingly being transferred directly to wives and children, particularly after the land has been planted either wholly or partially with cocoa trees. Such transfers are termed "gifts" and individual rights on such land are firmly established.[4]

In the communities studied, most migrants lease or hire land from customary landowners, or enter share tenancy or sharecropping agreements. Early migrant settlers in the area were able to acquire large tracts of land cheaply. However, later migrants have been faced with more developed land markets and growing land scarcity. The majority therefore depend on share tenancy for access to land (see for example Boadu 1992). As a result, 28.2 percent of the migrants interviewed for this study did not own land and depended on land hiring and sharecropping, whereas only 7 percent of the indigenous people were in this situation (Table 6.2).

Shared tenancy/sharecropping

Two main systems of share tenancy or sharecropping arrangements occur in this area: the *abusa* system (where the tenant receives a third of the cocoa output as his wages) and the *abunu* system (50:50 sharing arrangement).

In the *abusa* system, the tenant is admitted onto an already existing cocoa farm and becomes primarily responsible for maintaining and harvesting the crop. In the *abunu* system, by contrast, tenants are required to plant cocoa trees usually on bushland managing the trees until the whole field is planted with cocoa, at which time land ownership, rather than output, is usually divided between the tenant and the landowner on a 50:50 ratio (see Asenso-Okyere *et al.* 1993). In all the villages tenants normally have restricted rights to the use of land they have acquired. The tenant farmer is not free to cultivate any crop he likes without the consent of the landowner (see for example Benneh 1988). The land given in ownership cannot be sold, sub-let, or mortgaged. The land reverts to the landlord upon the death of the tenant, but, if the landlord does not need it, the children of the deceased may carry on farming it.

Share tenancy arrangements in Ghana differ from one locality to another. In most places these systems have undergone transformations over time. Amanor (1994) provides evidence of how the *abusa* system has changed over the years into the *abunu* system. He notes that in the period 1920–40, landowners appropriated a third share of the proceeds of the cocoa harvest or received a third portion of the cocoa farm created under the *abusa* tenancy. However, in the post-war period, as land became scarce and expensive, this arrangement was increasingly transformed into a half share (*abunu*) tenancy system.

The fact that share tenancy is widely practiced in this area with the ultimate division of land ownership occurring between contracting parties, and because cocoa tree planting is a long-term investment (lasting more than 70 years), rights appear to remain in perpetuity. This suggests that land ownership rights have been strongly individualized so as to allow "permanent alienation" of land (Kasanga and Kotey 2001).

Land purchase

At present, land purchase (outright alienation of land) is rare in this area, as Ghana's constitutional provision regulating the land market prevents the allocation of land on a freehold basis (Larbi 1998). People who are not part of the land owning group hold land on a renewable leasehold basis, usually for a long period (50–99 years). The local people, however, refer to such leases as "land purchase". Interviews with farmers suggest that strong land-rights are conferred on individuals who have acquired land on a leasehold basis.

Although land purchase is rare in most areas of Ghana, several authors maintain that cases of land purchase have occurred in the past (particularly in the Eastern region). This was in fact one of the commoner means by which earlier migrant farmers gained access to land (Hill 1963; Amanor 1994; Knox 1998; Amanor and Diderutuah 2001; Kasanga and Kotey 2001).

For example, Amanor (1994) asserts that the expansion of demand for palm oil during the mid-nineteenth century intensified pioneer frontier colonization. As a result, land purchase became predominant in the south-east forest ecotone. Estates that had gained large profits from oil-palm production offered compensation for expropriated land and expressed a desire to buy any surplus land offered to them. Vendors were found throughout the forest interior of Akyem in the Eastern Region. This area was still suffering from the repercussions of centuries of war, reflected in economic decline and poor transportation. The population was unable to take advantage of the opportunities to reorganize agricultural export production. As a consequence, the chiefs sought to gain profit from the export boom by selling land to migrant farmers rather than developing export crop estates.

Knox (1998) also notes that the introduction of cocoa to Ghana in 1879 provoked ıadical tenure changes as crop profitability lured farmers to the Akyem Abuakwa area in the Eastern region and other cocoa-growing areas of southern Ghana. Since profitability rather than subsistence was their purpose, local leaders rationalized that the land ought to be sold to them instead of granted, as had been the traditional means of allocating land to outsiders. Two main forms of tenure took root. On one hand, unrelated groups of individuals such as the Krobo purchased land and received parcel strips in accordance with their contribution to the purchase. On the other hand, the Akan purchased land, individually or among extended families, often in tracts larger than they were capable of cultivating themselves as land became viewed as an investment (Knox 1998). Hill (1963) also observed that the chiefs were glad to seize the opportunity of selling land outright to enterprising migrant cocoa farmers. It remains, however, unclear as to whether such "land purchases" constitute a final alienation of land from the chiefs' control.

In the Wassa Amenfi District, a significant controversy has occurred between landowners and migrant farmers over whether such "land purchases" give absolute title to the land. Several of the chiefs refused to acknowledge such transactions as constituting a final alienation of land. As the paramount chief of the Wassa Akropong traditional area pointed out:

> These tenant farmers fail to come to me for introduction and for confirmation of the transactions entered into. Furthermore, very few, if any, of the tenant farmers have valid documents or titles to the lands they have acquired. Some chiefs issue temporary receipts, which serve as documents on the land, while other tenant farmers make site plans, which only the chiefs sign. No indentures are prepared for the signatures of the divisional chiefs and their elders as well as

> mine. In the absence of indentures and accompanying survey papers, disputes concerning boundaries are legion (Bassayin, quoted in Benneh 1988: 233).

Land holdings

As the district's population has increased, land holdings have tended to decrease in size and the fallow area, still important for restoring fertility, has been reduced. Holdings ranged between 0.5 and 40 ha with an average land size of 6 ha. The majority of respondents (60 percent) used up to 5 ha of land with 25.6 percent utilizing between 6 and 10 ha. Only 1.3 percent of the respondents occupied over 25 ha of land (Table 6.3). Often these holdings are further divided among household members, leading to the progressive fragmentation of the farm.

Land holdings show similar patterns of distribution for both indigenous people and migrants, and for both men and women. However, migrants, and particularly early migrants, tend to have larger land holdings than indigenous people. For example, 28.2 percent of the migrants owned between 6 and 10 ha as opposed to 18.6 percent of the indigenous people, and all the respondents who owned more than 20 ha were migrants (Table 6.4). These results corroborate earlier findings by Amanor (1996) who noted that land owned by migrant men and women in this area tended to be of larger area than land owned by locals. Migrants both have longer traditions of commercial cocoa farming in this region and were instrumental in opening up the district for cocoa cultivation. By the time local farmers turned to cocoa farming, the new arrivals had already appropriated most of the land. As Benneh (1988: 231) notes:

> The native Wassa were slow in taking up cocoa cultivation; they were, however, interested in granting land to strangers in return for cash or other benefits. This was done through encroachment on virgin forestland. Anyone clearing an area of land, which has not before been cultivated, established his right over that tract of land. This method of land acquisition led to ecologically damaging practices. People cleared large tracts of land, felled the trees, and left the land uncultivated, merely in order to establish their rights over it. The holders of such large tracts now lease portions to land-hungry Wassa and migrant farmers...

Frontier development in Ghana's forest zone has resulted in complex social relations of production and systems of land tenure. This phenomenon has been a significant factor in the alienation of land and labor and has facilitated processes of social differentiation (Amanor 1994). This has led

Table 6.3. Size of land holdings in the study area

Land area (ha)	Number of respondents	% respondents	Sureso (%)	Kamaso (%)	Kamaboi (%)
≤ 5	96	60.0	72.3	61.8	37.5
6–10	41	25.6	18.5	23.6	40.0
11–15	14	8.8	6.2	7.3	15.0
16–20	5	3.1	1.5	3.6	5.0
21–25	2	1.3	1.5	0.0	2.5
25 ≥	2	1.3	0.0	3.6	0.0
Total number of respondents	160		65	55	40

Table 6.4. Size of land holdings in relation to migrant and indigenous people, and gender (percent of respondents)

Land area (ha)	Indigenes (%)	Migrants (%)	Men (%)	Women (%)	Total (%)
≤ 5	74.4	54.7	56.1	66.1	60.0
6–10	18.6	28.2	26.5	24.2	25.6
11–15	4.7	10.3	12.2	3.2	8.8
16–20	2.3	3.4	2.0	4.8	3.1
21–25	0.0	1.7	1.0	1.6	1.3
25 ≥	0.0	1.7	2.0	0.0	1.3
Total number of respondents	43	117	98	62	160

to the emergence of powerful landlords, absentee landowners with multiple holdings, and landless laborers and smallholder cultivators with insufficient land to farm.

Agricultural production and household food security

Agriculture dominates the economy of the entire Wassa Amenfi District, absorbing almost 85.6 percent of the district's working population (WADA 1996). More than 88 percent of the survey respondents reported that they cultivate cocoa. Indeed, relative prosperity in the study villages is reflected by the extent of cocoa farming. Other important cash crops are oil-palm, kola, and coconut.

In all the villages studied the majority of households relied on farming for the bulk of their foodstuffs. Cassava, plantain, yam, maize, rice, and vegetables are the major food crops cultivated in this area. In all the communities no farmer used fertilizers. Soil fertility is therefore maintained through a bush fallow system. However, fallow periods have shortened considerably due to land shortages resulting from land fragmentation occurring due to the influx of farmers into the new cocoa pioneer frontier. Indeed, the majority of farmers maintained that fallow periods have declined to only two to four years. Most complained that their land holdings are so small that they are unable to fallow land properly, if at all.

Farmers practice mixed cropping, where two or more crops are cultivated on the same piece of land in the same cropping season. Mixed cropping helps them to obtain a wide variety of produce and also provides a degree of security since one crop may succeed even if others fail. Mixed cropping follows a definite pattern in this area. Following slashing and burning, maize is the first crop planted, followed by cassava, and then plantain. Cocoa is planted after the maize is harvested. After harvesting the cassava, the plantains provide shade for the cocoa until eventually they themselves are harvested, leaving only the cocoa. The farm then becomes a cocoa plantation. The farmer then moves to another area to repeat the practice, clearing the forest in the process. Through this process farmers have gradually turned most of their land to cocoa farming, leaving very little space for food production. Consequently, many households are unable to produce enough food to feed their families.

Food insecurity is a major problem for all the study communities. As many as 61 percent of respondents indicated that the previous year's harvest did not provide enough food for the needs of their families. Most farmers reported that the problem of food insecurity in the area has come about due to land shortage resulting from population pressure and an over-reliance on cocoa production. As one woman from Kamaso Village recounted:

> All the land is finished...We do not have enough land for growing food crops. Food shortage is therefore a major problem in this village. Sometimes we have to travel to the city to buy food, which should not be the case. The forestland, which is fertile, is now finished and since our numbers are increasing fallow periods have reduced to between three and six years, by which time the soil had not regained its fertility. If the government does not help us, we will all leave for the city because we are dying of hunger. We are pleading with the government to release some of the forest reserve to us for farming (female interviewee, Kamaso, March 2001).

In a study of farming systems in this area, Amanor (1996) found that of a total farming area of about 2,630 acres estimated by 142 farmers to be within their farming system, 30 percent was a fallow (bush) area, 50 percent was a cocoa plantation, and 1 percent was other plantation crops, with food crop area constituting only 19 percent of the total farming area. He noted that the majority of farmers have less than 5 percent of their land under food crops, and between 20–60 percent under cocoa.

This trend has significantly shortened fallow periods and has led to a decrease in the yield of food crops. Meanwhile, farmers complain that the yield of cocoa has also declined considerably over the years since they cannot afford the necessary inputs such as pesticides and fertilizers. A reduction in the yield of cocoa has significantly reduced household income and therefore the ability of households to purchase food.

Dependence on forests and forest resources

As pressures on the agricultural land-base increase, leading to progressive fragmentation of farm holdings and the overuse of arable land, the ability of farm households to achieve self-sufficiency and livelihood security from their land decreases. Rural populations in these areas, especially the landless and later-stage migrants, are therefore becoming increasingly reliant upon the remaining forest resources for their livelihoods. As uncultivated forestland outside reserves has largely disappeared from this area, reserved forests have come under increasing pressure from local people.

Indeed, land tenure and land availability partly shape and influence the way rural people use and depend on forest resources, thus having a significant effect on the management of forest resources on which rural livelihoods depend. Many studies have indicated that, where people have limited land for agricultural production, forest foods and income from forest products often plays an important role in rural livelihoods.

Local people's dependence on forest resources derives from a number of sources: forest clearance; poverty—which increases dependence on minor forest products (for example, snail and mushroom gathering and the collection of Marantaceae leaves to wrap cooked foodstuffs); and vulnerability. Non-timber forest products (NTFPs) provide an important buffer or safety net in times of stress and when laboring work is in short supply (Agyemang 1996; Brown 1999).

Forests contribute to all aspects of rural life in the communities studied, providing food, medicines, firewood, fodder, canes, building materials (poles and thatch), wrapping leaves, pestles, chewing sticks, and materials for all sorts of household items, as well as many more intangible benefits such as cultural symbols, ritual artifacts, and sacred sites. NTFPs thus form an integral part of the rural economy.

Virtually all the respondents (98 percent) reported that they consume several varieties of forest foods in their households including fruits, nuts, seeds, leaves, mushrooms, snails, honey, tubers, and bush-meat. For example, when asked to mention the most important means by which they were able to survive the insufficient food harvest the previous year, the majority of respondents (48.1 percent) said they bought food, while 36.9 percent collected food from the forest or bush (Table 6.5).

These results show the important role forest foods play in the subsistence economy in the area, providing a safety net which people can fall on in times of need. In addition to forest foods, most household items such as baskets, sleeping mats, crop-drying mats, grinders, mortars, spoons, and other utensils are also produced from forest resources found in the surrounding environment (see Photo 6.1). Forest resources thus feature prominently in the material culture of the study villages.

Although a significant proportion of forest products are used in the household, the sale of gathered forest products and the processed goods derived from them also provide an important contribution to the family income throughout the year. In the study area, forest-based activities are one of the most common income-earning options for households, although the majority of the rural people are farmers for whom forest-based activities represent a supplementary source of income. In all, 94 percent of households obtain some sort of income from forest product activities (see Photo 6.2). These activities are particularly important at times when agricultural tasks diminish, or during acute financial need.

The returns on labor invested in NTFP activities are often very low. Hence such work is most popular among those with the fewest alternatives—people in situations of stress for whom any level of return may be better than none.

Table 6.5. Means of surviving insufficient food crop harvest (percent of respondents)

Means	% respondents
Borrowed food	4.4
Bought food	48.1
Given food	9.4
Collected food from the forest/bush	36.9
Other ways	1.3
Total number of respondents	160

Photo 6.1. The use of canes for making baskets—the weaving of cane baskets is an important source of household income in the study villages

Photo 6.2. Men engaged in wood carving activities—a wide variety of household items are carved from wood, providing a source of income for many people

These are predominantly the landless and recent migrants, particularly those from the north. Indeed, households in the latter category tend to be far more dependent on forest-based activities than the indigenous households. For example, 96.6 percent of the migrant households reported that they obtain income from forest-based activities compared with 88.4 percent of the indigenous households.

Migration of individuals, principally to establish cocoa farms, is a significant factor in the development of agriculture, also having consequences for the level of commercialized forest-based activities in the study area. In the earlier stages of establishing a cocoa farm, migrant farmers supplement their incomes by engaging in forest-based activities. As mentioned earlier, later-stage migrants are confronted with more developed land markets and growing land scarcity and therefore have less land and fewer income options. The majority therefore depend to a greater extent on forest-based activities for their initial livelihood.

Forest degradation and decline

Pressure from local peoples on the forest in terms of forest clearance and forest product extraction has significantly worsened forest conditions and led to a decline in resource availability. In order to gain a general picture of local peoples' perceptions, households were asked to describe the condition of the forests, especially those in forest reserves. Virtually all respondents maintained that the condition of the forests was very poor, due to widespread degradation. Most described the reserved forests as "government grassland" due to the lack of tree-cover. Others also described them as "deserts". Although the majority of people admitted that farming has negatively modified the forest ecosystem, others attributed the degradation of forests to the activities of logging companies and chainsaw operators. Most respondents were worried that their source of livelihood was being eroded. They reported that the availability of most NTFPs has declined considerably, with some products becoming almost extinct.

Discussions with key informants suggest that forest degradation and the decline of forest products in the area is closely associated with both timber extraction and population pressure. These discussions also reveal that since the 1950s each decade has seen a major change in the availability of forest products. Most believe that forest products began to decline in the 1960s as logging became intensive in the area and loggers opened up a significant proportion of the forests. Respondents maintain that although logging continued well into the 1970s, the most important driver of forest decline during the 1970s was the migration of tenant cocoa farmers into the area, as forest areas were cleared and turned into cocoa plantations. Land became fragmented as the population of tenant farmers increased. Pressure on forestlands and forest resources also increased, leading to unsustainable harvesting practices.

Several respondents claim that by the late 1980s virgin forests outside reserves had virtually disappeared. Certain NTFPs had become scarce and some were almost extinct. Urban-based commercial gatherers began harvesting NTFPs in the early 1990s, further aggravating the problem. Today, certain species providing bush-meat, such as the red colobus monkey, are reported to have almost disappeared from the area.

In a recent major study of Ghana's forest composition and botany, Hawthorne and Abu-Juam (1995) found considerable degradation in the HFZ and reported that only about half of the nation's 1.8 million ha of forest reserves was in reasonable condition. They noted that approximately 14 percent of the reserves have no significant forest left and that there is virtually no substantial area of forest outside the reserves (Table 6.6). This fact is also recognized by the Ghana Forest Service Division, which estimates that there is very little intact forest currently left outside the forest reserves, wildlife reserves, and traditionally protected areas of Ghana (Adu and Owusu 1996).

Although local communities have played a major role in opening up the forests of Ghana for agricultural production, farming is not the only cause of forest decline and local people are not the only agents of forest degradation. Several authors maintain that in addition to agriculture and logging, other factors such as firewood collection, uncontrolled wildfires, mining, and urban growth are immediate and direct causes of forest decline and degradation in Ghana (Adu and Owusu 1996; Britwum 1989; Hawthorne and Abu-Juam 1995; Mery 1996). Beyond the immediate causes lies a complex mix of intersecting social, economic, cultural, and political conditions, which aggravate many of the immediate factors and make the solution of the deforestation problem more difficult. These include: uncertain forest property rights; high population growth rates which accentuate the effects of most of the immediate causes; lack of sustainable agriculture technologies; and macro-economic policies and socioeconomic factors that lead to stressed living conditions for vulnerable groups and internal migration to frontier agricultural areas (Adu and Owusu 1996).

Table 6.6. The condition of Ghana's forest reserves

Condition	Area (km^2)	% total
Excellent	346	1.9
Good to OK	2,485	14.0
Partly degraded	6,190	34.9
Mostly degraded	3,697	20.8
Very bad	2,626	14.8
No forest	2,401	13.5
Total	17,745	100

Source: Derived from Hawthorne and Abu-Juam (1995: 15).

Conclusions

In this chapter I analyzed migration into the Western Region of Ghana, one of the remaining forest frontiers of the country. Commercial cocoa production has been a key element in this process, influencing the impact that immigrants have had on various aspects of rural life in the forest frontier zone. However, cocoa production is only one factor that explains the impact of migration on this region. The cultural makeup of locals and migrants, local biophysical conditions, government economic policies, and world markets for cocoa and other products all contribute to the understanding of the processes described in this chapter.

In the chapter I examined in detail the outcomes of migrating groups that came to the Western Region to find land for commercial cocoa production. This process has changed social relations of production and land tenure systems, alienating land and labor and facilitating the process of social differentiation (Amanor 1994). This migration has resulted in the emergence of large landlords, absentee landowners with multiple holdings, and landless labor and smallholder cultivators with insufficient land to farm.

Although some of the changes described here could also have been expected to occur in situations where internal growth resulted in increased pressure on resources, several changes can clearly be attributed to migration processes. These particularly relate to the changes of ethnic composition at the forest frontier, distribution of land ownership, evolving tenure systems, and variations in resource use patterns of current residents.

The forest frontier populations in the Western Region are faced with a daunting task. Pioneer frontier colonization in Ghana is now in its final stages. Most of the remaining forests in the Western Region have already been converted for cocoa production (Amanor 1994). Confronted with the decline in frontier land, a crisis of production exists in many areas. Farmers will increasingly have to find solutions to mounting problems and devise new adaptive technologies and farming strategies.

Acknowledgements

This paper is based in part on material collected during field research in Ghana in 2001 for a doctoral thesis at the University of Hull, UK. I am grateful to my supervisor, Dr. Phillip Bradley, for his comments on an earlier draft of this paper. I would like to acknowledge the financial support of the Japan Center for Area Studies, the Forestry and Forest Products Research Institute, and the Center for International Forestry Research, along with the

Center for Southeast Asian Studies, Kyoto University, Research Institute for Humanity and Nature, and the National Institute for Environmental Studies. Finally, I am grateful to the organizers of the symposium: "In, Out, In, Out: Populations, Migrations, and the Social Ecology of Tropical Forests," (Kyoto December 2002) especially Abe Ken-ichi, Wil de Jong, and Lye Tuck-Po for inviting me to participate.

Notes

1. Administratively, Ghana is divided into ten regions: Ashanti, Brong-Ahafo, Central, Eastern, Greater Accra, Northern, Upper East, Upper West, Volta, and Western Regions.
2. Apart from administrative districts, the whole of Ghana is divided into approximately fifty forest districts. The Asankrangwa forest district is one of the largest forest districts in Ghana.
3. These are 1984 population figures. Recent figures may be higher than these.
4. Such gifts are formalised by the husband presenting ritual drinks to family elders. This ritual, in which family elders and other members are present as witnesses, is crucial so that the transfer of land-rights by gift will not be contested in the future.

References

Adu, S.V. and J.G.K. Owusu. 1996. Natural resource degradation in Ghana: Causes and consequences. In F. Ofori and E.Y. Safo, eds., *Proceedings of the National Workshop on Soil Fertility Management Action Plan for Ghana held at Cape Coast, 2nd–5th July 1996*. Accra, Ghana: Ministry of Food and Agriculture.

Agyemang, M.M.O. 1996. The leaf gatherers of Kwapanin, Ghana. *Forest Participation Series No. 1*. London: International Institute for Environment and Development (IIED).

Amanor, K.S. 1994. *The New Frontier: Farmers' Response to Land Degradation—A West African Study*. Geneva: UNRISD, and London: Zed Press.

Amanor, K.S. 1996. *Management of Trees in Farming Systems: Report on Asankragwa, Akyem Oda and Mampong Ashanti of Ghana*. Legon, Ghana: Institute of African Studies, 1–61.

Amanor, K.S. 2000. Farmers, forestry and fractured environmentalisms in

Ghana's forest zones. In R. Cline-Cole and C. Madge, eds., *Contesting Forestry in West Africa*, 307–321. Aldershot: Ashgate.

Amanor, K.S., and M.K. Diderutuah. 2001. *Share Contracts in the Oil Palm and Citrus Belt of Ghana*. London: International Institute for Environment and Development.

Arhin, K. 1985. *The Expansion of Cocoa Production: The Working Conditions of Migrant Cocoa Farmers in the Central and Western Region*. Accra, Ghana: Mimeo.

Asenso-Okyere, W.K., S.Y. Atsu, and I.S. Obeng. 1993. *Communal Property Resources in Ghana: Policies and Prospects*. Discussion Paper 27. Legon, Ghana: Institute of Statistical, Social and Economic Research, University of Ghana.

Benneh, G. 1965. Land tenure and farming systems in Nkrankwanta: A village in the pioneer cocoa area in Ghana. *Bulletin of the Ghana Geographical Association*, 7(2): 6–15.

Benneh, G. 1988. The land tenure and agrarian system in the new cocoa frontier of Ghana: Wassa Akropong case study. In W. Manshard, and W.B. Morgan, eds., *Agricultural Expansion and Pioneer Settlements in the Humid Tropics*. Tokyo: The United Nations University.

Boadu, F. 1992. The efficiency of share contracts in Ghana's cocoa industry. *Journal of Development Studies*, 29: 108–120.

Britwum, S.P.K. 1989. Industrial forest plantations development in Ghana–problems and prospects. In J.G.K. Owusu, ed., *Revision of Ghana's Forest Policy*. Forestry Commission Symposia Series 3. Accra, Ghana: Forestry Commission.

Brown, D. 1999. *Principles and Practice of Forest Co-management: Evidence from West-Central Africa*. European Union Tropical Forestry Paper 2. London: ODI.

Ghana Statistical Service (GSS). 2000. *2000 Population and Housing Census: Provisional Results*. Accra, Ghana: Ghana Statistical Service.

Hawthorne, W.D. and M. Abu-Juam. 1995. *Forest Protection in Ghana with Particular Reference to Vegetation and Plant Species*. Gland: IUCN and Cambridge: ODA.

Hill, P. 1963. *The Migrant Cocoa Farmers of Southern Ghana: A Study in Rural Capitalism*. Cambridge: Cambridge University Press.

IUCN. 1992. *Protected Areas of the World: A Review of National Systems. Volume 3: Afrotropical*. IUCN, Gland, Switzerland and Cambridge

Johnson, M. 1964. Migrants in progress. *Bulletin of the Ghana Geographical Society*, 9(2): 4–27.

Kasanga, R.K. and N.A. Kotey. 2001. *Land Management in Ghana: Building on Tradition and Modernity*. London: International Institute for Environment and Development.

Knox, A. 1998. Ghana Country Profile. In J.W. Bruce, coordinator, *Country Profiles of Land Tenure: Africa, 1996*. Land Tenure Centre Research Paper 130. Madison: Land Tenure Centre, University of Wisconsin.

Larbi, W.O. 1998. Tenure transformation and land valorization processes at the urban periphery of Ghana. Paper presented at the International Conference: *Land Tenure in the Developing World with a Focus on Southern Africa*, University of Capetown, South Africa, 27–29 January, 1998.

Mery, G. 1996. Deforestation and sustainable forestry challenge in Ghana. In M. Palo and E. Yirdaw, eds., *Sustainable Forestry Challenges for Developing Countries*, Dordrecht: Kluwer Academic Publishers.

Nsenkyire, E.O. 1992. *Ghana—Progress towards the Year 2000: Sustainable Management of Tropical Forests*. Unpublished.

Prah, E.A. 1994. *Sustainable Management of the Tropical High Forest of Ghana*. London: Commonwealth Secretariat.

Quisumbing, A.R., E. Payongayong, J.B. Aidoo, and K. Otsuka. 1999. *Women's Land Rights in the Transition to Individualised Ownership: Implications for the Management of Tree Resources in Western Ghana*, Food Consumption and Nutrition Division Discussion Paper 58. Washington, DC: International Food Policy Research Institute.

Richards, P. 1985. *Indigenous Agricultural Revolution: Ecology and Food Production in West Africa*. London, Hutchinson and Boulder: Westview Press.

Robertson, A.F. 1987. *The Dynamics of Productive Relationships: African Share Contracts in Comparative Perspective*. Cambridge: Cambridge University Press.

Smith, E.K. 1994. Forestry as a land use issue in Ghana. Paper presented at the international workshop: *Forest Land Use Options: Conflicts and Solutions*, Kumasi, Ghana, 24–28 January 1994.

Smith, E.K. 1999. *Developments and Setbacks in Forest Conservation: The New Political Economy of Forest Resource Use in Southern Ghana*. Natural Resources Management Programme, Ministry of Lands and Forestry Technical Paper, Accra, Ghana, January 1999.

Wassa Amenfi District Assembly (WADA). 1996. *Wassa Amenfi District Development Plan: 5-year medium term plan: 1996–2000*. Draft Report. WADA, Asankrangwa, Ghana.

Wilks, I. 1977. Land, labor, capital and the forest system of Asante: A model of early change. In J. Friedman and M.J. Rowlands, eds., *The Evolution of Social Systems*. London: Duckworth.

World Bank. 1987. *Ghana Forestry Sector Review*. Washington, DC: World Bank

World Conservation Union (IUCN). 1992. *The Conservation Atlas of Tropical Forests: Africa*. London: Macmillan.

7
Spatial Shifts and Migration Time Scales among the Baka Pygmies of Cameroon and the Punan of Borneo

Edmond Dounias and Christian Leclerc

Introduction

Since the 1966 "Man the hunter" conference (Lee and DeVore 1968), the ecological perspective has become a pertinent tool for understanding mobility and migration among hunter-gatherer societies (Casimir and Rao 1992; Leacock and Lee 1982). The ecological adaptation of hunter-gatherers was still of interest to many papers presented at the Conference on Hunting and Gathering Societies in Edinburgh, September 2002. Although the ecological approach to human mobility has often been challenged (Barnard 1983; Bender and Morris 1988; Burnham and Ellen 1979; Ellen 1989; Ellen and Fukui 1996; Ingold *et al.* 1988; Kent 1996; Kent and Vierich 1989; Schrire 1984), this field of studies has contributed much to the elaboration of cultural ecological models and conceptual frameworks concerning the interactions between a given environment (the ecological system) and the populations that occupy it (the social system).

The ecological system approach—encompassing the ecosystem, the cognitive system, and the environmental exploitation system—has led to significant conceptual and empirical advances in the understanding of the mobility of hunter-gatherer societies and its admitted interference on landscape dynamics, especially through ethno-archeological studies (Binford 1980; Butzer 1982; Gremillion 1997; Hitchcock 1982; Hutterer 1983; Ichikawa 1999; Laden 1992). Such an approach is particularly useful in the analysis of human–forest interactions at a very local level during a limited period of time. However, social ecology systems are insufficient when considering the following: migration on a broader spatial (for example, regional) and temporal (for example, a century or more) scale; and the long-term conservation of forested areas. The environment affects individuals and individual social relations. The factors determining how humans modify the landscape today necessitate historical contextualization (Balée 1998; Boomgaard 1997, 1998; Crumley 1993; King 1976). As stressed by Eghenter (1999; Chapter 11), migration literature pays too much attention to economic motivations and often ignores other pertinent historical factors such as warfare, intra-village disputes, religious beliefs, and demographic and environmental changes. The historical approach provides a good safeguard against deterministic interpretations of the ecological systems approach.

The interactions observable today between forest dwellers and their forest geography is strongly influenced by political events that sometimes occurred several decades or even centuries ago (King 1976). We seek to illustrate this political aspect by considering the historical migration of two groups of forest foragers from two geographic areas: the Baka Pygmies of Southern Cameroon in Africa (Figure 7.1), and the Punan of Eastern Kalimantan in Southeast Asia (Figure 7.2). We chose these groups because they share a hunter-gatherer background. They have also changed in similar ways. Many subgroups have become true swiddeners, while others have left the forest for off-farm activities (Biesbrouck *et al.* 1999; Levang *et al.* in press). Many have substituted traditional and exclusive relationships with particular groups of villagers (their historical patrons) for new alliances. These forest dwellers now settle for longer periods in villages. Nevertheless, they continue to spend several weeks or months at a time in the forest for hunting and gathering purposes, although production for commercial markets now occupies more attention than production for domestic consumption. Although the historical events that have affected each group are completely different, they similarly and severely influenced the current geographic distribution of the groups. In the cases of the Baka and Punan we refer to analyses offered by numerous contributors describing how historical events have changed these two hunter-gatherer societies.

Comparative historical linguistics supplemented by historical, ethnographic, and archival data emphasize how the slave trade in the Guinean Gulf and the control of forest product trade by Borneo's coastal kingdoms shaped, at a regional level, the residential and linguistic distribution of the forest communities observable today. Therefore, we argue that the political context orienting the migration of human groups at a regional scale has no necessary relation to the social and ecological factors determining mobility at a local level. However, we do not pretend that the historical approach obliterates that of the socio-ecological. Each approach definitely complements the other. We distinguish between them so as to more clearly show that effects are intermingled regarding the concrete issue of forest management. We advocate that development actors, researchers, and decision makers, who

Figure 7.1. Pygmy groups of Congo Basin and Baka distribution

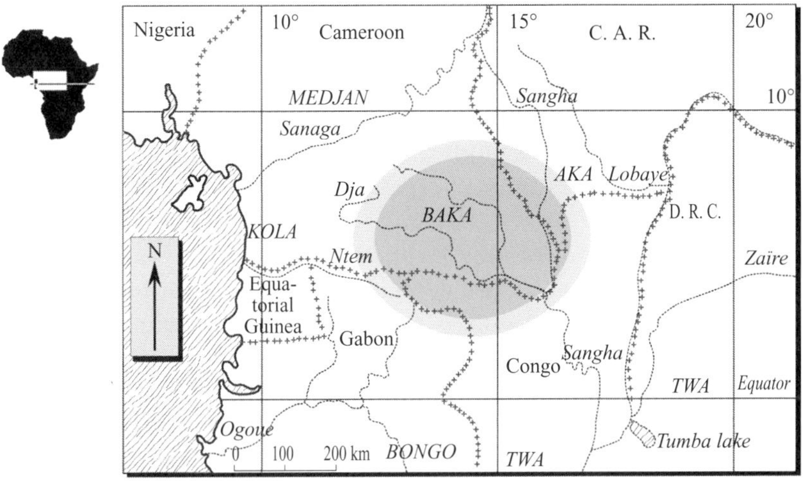

Source: Dounias (2001).

Figure 7.2. Punan groups of the Malinau watershed and neighboring areas

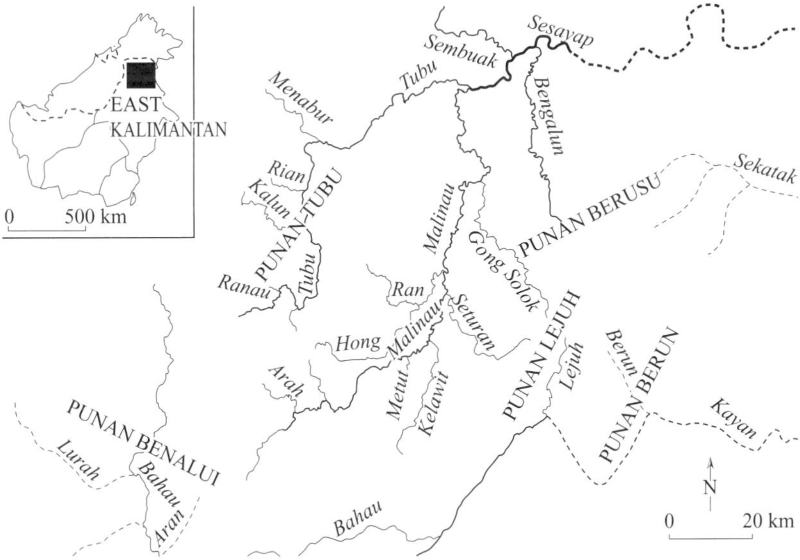

Source: (Wahyantono 2001).

necessarily intervene at very circumscribed spatial and temporal scales, must take macro-events into account. Such attention to history may improve the understanding of local strategies, especially in the crucial issues of land tenure and access to resources.

The slave trade and its implications for tropical forest migration

Social and ecological features of contemporary Baka

The study of the Baka's current demography and distribution is facilitated due to the fact that they live alongside main roads together with their farmer allies (Figure 7.3). Using the census conducted by Dhellemmes (1977–1982), we estimate a Baka population of 26,000 living in 410 villages along the roads. Less than 13 percent of this population is located in the Pahouin distribution area. According to oral tradition, this group moved there as a result of migration during the twentieth century. The highest density of Baka

Figure 7.3. The location of Baka villages along roads

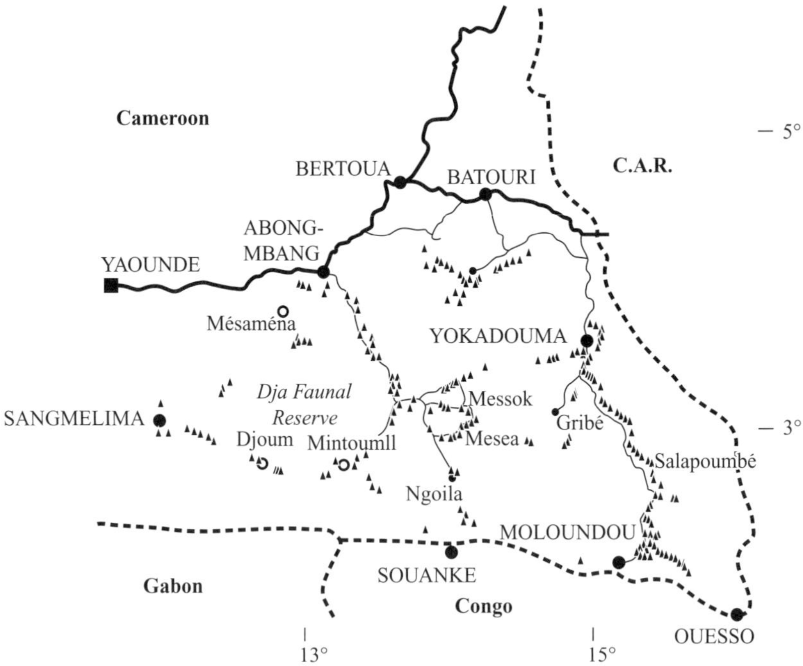

Source: Leclerc (2001).

Figure 7.4. Baka distribution in Cameroon departments

Source: Adapted from Dhellemmes (1977–1982).

people appears in the most southeastern part of Cameroon (Figure 7.4) in what used to be a refugee area for populations threatened by slave traders.

The Baka basic socioeconomic unit is the camp consisting of an average of seven families (Bahuchet 1985; Bahuchet and Guillaume 1982). The camp is constituted around a nucleus of parents, spouses, and the parents of spouses. Every Baka belongs to a patrilineal descent group, but the camp includes members of several lineages. Local groups move along a specific trail, and several camps may follow the same trail. Neighboring camps have ongoing exchange relationships, and members may visit each other daily when camps are settled close together. On many occasions, conjugal families move to visit relatives in other camps. Such visits can last from a few days to several months. Patrilineages are distributed across the ethnic group's area of occupation; no group owns any particular territory. Through the dispersion of relatives and the extension of relationships through conjugal ties, each Pygmy has right of access to several territories in addition to his or her birth-territory (Bahuchet 1992).

Since the middle of the twentieth century, the Baka have established permanent villages along main roads in close proximity to their farmer neighbors. They have adopted shifting cultivation but continue to migrate into the forest on a seasonal basis. In fact, the Baka spend on average only 28 percent of their time in the village (Figure 7.5). The highest ratio of presence in the villages corresponds to the long dry season when the demand for farm labor is most intense. Mobility between villages is infrequent but motivates the longest absences. Social events (such as visits to relatives and attendance at funerals) are the primary causes of inter-village movement (Leclerc 2001).

The differentiation between the Baka and the Aka

Several valuable studies have illuminated historical events in the Congo River Basin with particular emphasis on the humid forest zone of Cameroon acknowledging the deep historical articulations of diverse peoples throughout

Figure 7.5. Annual rainfall and Baka presence in Messea village

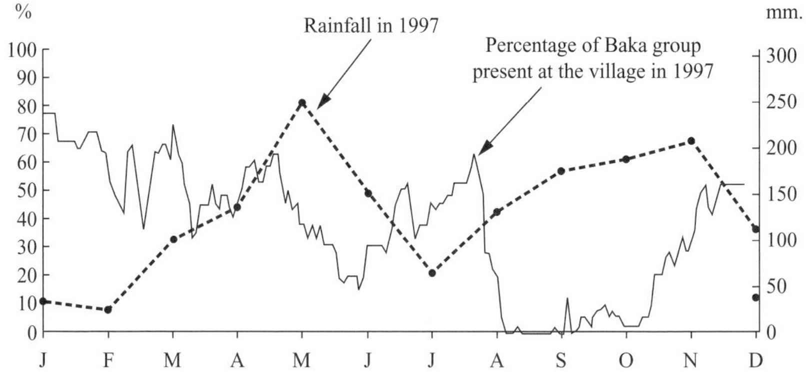

Source: Leclerc (2000).

Equatorial Africa (Coquery-Vidrovitch 1972; Harms 1981; Tardits 1981; Vansina 1990). Ethnolinguistics has become an important tool contributing to the historicization of hunter-gatherers in Central Africa, particularly offering reconstructions of social relationships to foster an understanding of patterns of technological, social, and linguistic exchange among diverse forest communities (Bahuchet 1993a, 1993b; Monino 1995).

However, there is a strong temporal discrepancy between the histories of languages and language speakers that informs us about the ancient Bantu population and more local movements respectively (Leclerc 2001). We propose to describe the events that affected Baka migration during the nineteenth century. First, it is necessary to refer to older historical events that account for a common origin between the now distinct Baka and the Aka Pygmies of the Central African Republic. Bahuchet (1993a, 1993b) demonstrated this common origin through an extensive analysis of vocabulary and social and cultural practices among a wide range of Central Africa forest dwellers. He has called *Baakaa the common ancestral population of the Baka and Aka. The livelihood of the *Baakaa did not differ from that of the current Aka and Baka groups, although both populations moved extensively throughout the Central African rainforest adapting their languages along the way. Bahuchet estimates that the *Baakaa divided during the sixteenth century when they came into contact with migrating Oubanguian and the Bantu expansion.

Around 1600, the ancestral population of the Baka and Aka split into two groups: one lived with the Oubanguians and the other with Bantu speakers. Over the course of the following two to four hundred years, these groups moved from east to west with their new farming partners. Pygmies and their Oubanguian and Bantu neighbors were in constant contact until the period 1620–1650. During these migrations, the Pygmies provided their allies with raffia, lumber, magical products, and bush-meat and were active protagonists of the Great Congolese Trade by providing ivory and padouk red wood (*Pterocarpus spp.*). Between 1650 and 1690, the two groups—Oubanguian/Baka and Bantu/Aka—separately contributed to this trade. By around 1880, before the arrival of the Germans and before the colonial abolition of slavery, Aka and Baka had completed their individualization process and become distinct groups occupying their current areas of distribution (Figure 7.1).

By the end of the eighteenth century, the Oubanguian-speaking group fragmented and dispersed, leading to the linguistic isolation of subgroups. The extended spatial dispersal of current Oubanguian speakers is revealed by the high diversity of Oubanguian languages now spoken throughout a large area. Only a wave of panic could explain such fragmentation, and the only possible cause of this terror could have been the slave trade. Today the

Baka—who speak an Oubanguian language inherited from their close ties with Oubanguian speakers prior to fragmentation—now entertain relations with sixteen different ethnic groups, of whom fourteen are Bantu speakers. These groups moved into the forest with the Baka as guides, escaping from the Nzimo slave raiders.

The Baka in slave raiding times

Central African hunter-gatherers participated in multiple trade networks, retaining a strong degree of autonomy until the era of Atlantic Trade roughly between 1500 and 1700. During this period violence and competition for resources increased. This led to a loss of independence for many agriculturalists and hunter-gatherers alike, as some agriculturalists began raiding other communities forcing them into slavery. According to Klieman (1999), there is no evidence of subordinate social status among hunter-gatherers until the era of Atlantic Trade, when different ethnic groups entered into complex systems of clientelism, pawnage, and slavery, either through coercion or expediently for survival purposes.

Figure 7.6. Slave trade on the Loango Coast from 1637 to 1868

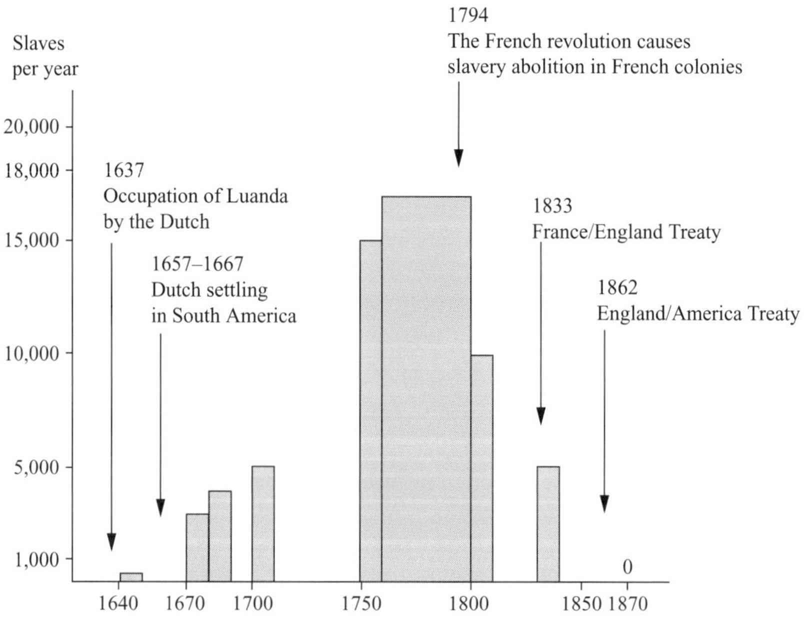

Source: Adapted from Bahuchet (1985).

The Dutch arrived in Central Africa and successively commenced exploiting and trading forest products, ivory, and copper. Their interest in slaves began around 1635 with the acquisition of new land in Brazil. This trade was initially limited to a few hundred slaves per year but its volume grew rapidly, reaching some 18,000 victims per year for the three harbors of the Loango coast alone (currently Congo, Figure 7.6) between 1765 and 1790 (Bahuchet 1985, 1993a; Martin 1948).

In Cameroon, the slave trade that reached its peak between 1850 and 1880 was organized along two main axes. The Fulani predominantly managed the northern axis, while the coastal Bantu tribes controlled the western route. The Fulani reached the limits of their expansion—the margins of the southern forest—around the period between 1860 and 1870 (Mohammadou 1978). They forged alliances with the Mkako and the Gbaya ethnic groups, tracking down potential slaves farther south into the forest, and pushing the northern Bangando-Mpomam to seek refuge deeper in the south (Rupp 2002). The forest served as a natural refuge as the Fulani cavalry was totally inefficient in forested lands. The active role of the Mkako and the Gbaya in tracking slaves is clearly demonstrated by the progressive integration of these two groups during a period when most eastern Cameroon populations were dispersing. The reported common chieftainship organization of the Mkako and Gbaya (Burnham *et al.* 1986; Copet 1981; Copet-Rougier 1987) can certainly be attributed to their collaboration during war-time and in the acquisition of slaves for the Fulanis. The Nzimo, the Badjue, and the Bulu largely took over the western slave-trade axis. This trade brought firearms and gun powder to the interior of the continent, and ivory and slaves to the coast (Figure 7.7).

A third indirect axis originated farther south, from three harbors of the Loango coast. The establishment of the slave trade from Congo and its spread inland following the navigable rivers has had a major impact on the distribution of forest dwellers in the interior, and has caused massive migration flows from south to north. The Bantu and the Oubanguian-speaking groups tried to seek refuge from slave traders by progressing farther north and hiding in near-inaccessible swampy areas. The far southeastern part of Cameroon thus appears to be a converging area for forest dwellers threatened by the slave trade.

The end of the eighteenth century epitomizes the peak of the slave trade along the Guinean coast. During that time, the French were the main traders of slaves. In just over two hundred years, from 1640 to 1868, nearly 600,000 slaves left Africa from the three harbors on the Loango coast, of whom nearly 10 percent died during the Atlantic crossing. These lives were exchanged for

Figure 7.7. Slave trade in Eastern Cameroon between 1850 and 1880

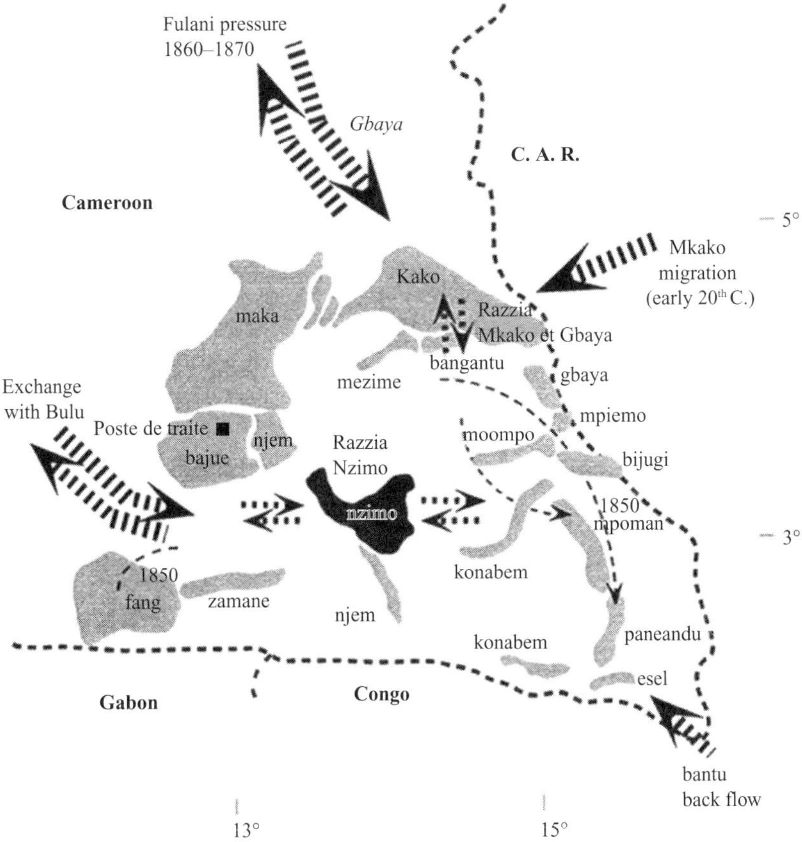

Source: Leclerc (2001).

a few million meters of fabric, 1.5 million firearms, several thousand tons of iron, 2.5 million knives and machetes, 5 million liters of alcohol, with an unexpected bonus of contagious and fatal diseases (Soret 1978). Eventually, ivory and palm oil replaced humans as the chief commodity (Figure 7.8).

Europeans finally entered the Congo Basin area in 1890. A decree published in 1899 officially divided the area into concessions granted to companies to exploit natural products: rubber, resin, oleaginous plants, ivory, and pelts. The same companies were also awarded commercial monopolies (Bahuchet and Guillaume 1982; Coquery-Vidrovitch 1972; Guillaume 2000). Rubber, ivory, pelts, and palm nuts were permanently in demand, while copal resin, raffia, rattan, and kola nuts were only occasionally traded.

Past slave trade and current Baka distribution

The current distribution of the Baka clearly reflects the movements of a population that fled from slave raiders. The Baka groups are primarily concentrated in the southeastern province of Cameroon. Their density declines at the periphery and they do not live in areas occupied by the neighboring Pahouin (Figure 7.8). Today, the Baka have minimal contact with any of the ethnic groups that actively contributed to slave trade, among them some Beti and Bulu clans, Vutte and Bamum clans, and other groups such as the Bassa who ensured the transit of slaves to the coast (Alexandre and Binet 1958; Geschiere 1981; Laburthe-Tolra 1981; Mohammadou 1991; Tardits 1980). Slave traders also had allies among the Pahouin and the Eastern Maka/Njem. The Pahouin and Maka/Njem are not truly homogenous groups. With the risk of oversimplification, the areas occupied by these two groups corresponds with the dividing line between active slave traders (the Beti and many assimilated groups) and the ethnic groups that were raided (mainly Oubanguian native groups of Eastern Cameroon) and concomitantly with the limits of the area where the Baka live. A few Beti and Bulu clans experienced a common form of domestic slavery (Alexandre 1965) and continue to exhibit

Figure 7.8. Pahouin and Maka-Njem groups

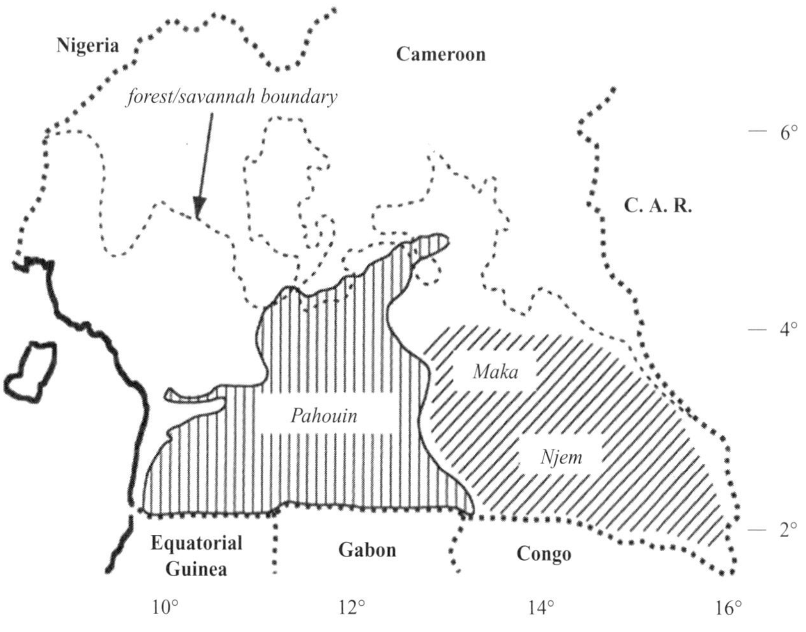

Source: Leclerc (2001), adapted from Alexandre and Binet (1958).

signs of stratified social organization. Such hierarchical structure persists among the Eton and Mangisa who still distinguish between the descendants of noble families and those of slave origins.

The oral traditions of almost all Oubanguian-speaking groups confirm a convergence on the southeastern part of Cameroon, indicating that the Baka guided the populations who tried to escape slave raiders. During the nineteenth century, this region experienced abundant population movement, contributing to the ethnic fragmentation observable today in southeastern Cameroon (Figure 7.9). Current linguistic geography and the selective absence of socio-cultural interactions with certain Pahouin groups contradict suggestions that some Baka groups served as guides to slave raiding societies. In fact, the evidence indicates that all the Baka groups assisted the raided populations to find refuge in the less accessible parts of the forest.

Figure 7.9. Distribution of linguistic groups (other than Baka) in Eastern Cameroon

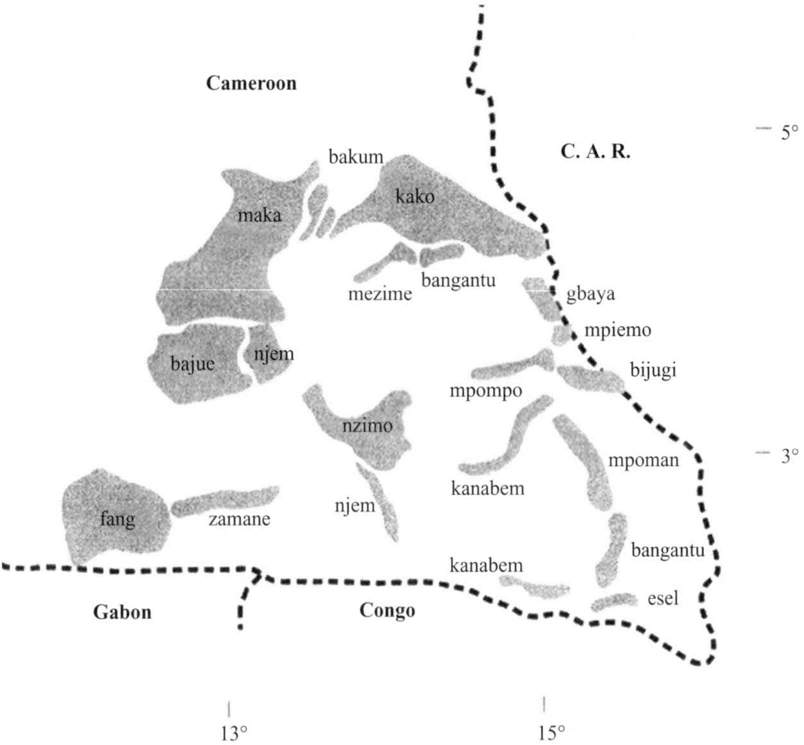

Source: Leclerc (2001), adapted from Dieu and Renaud (1983).

Trade in forest products and migrations of tropical forest peoples

The Punan: A complex patchwork of distinct groups

The Indonesian province of East Kalimantan is home to some 10,000 Punan hunter-gatherers, scattered all over the province in small hamlets housing on average ten to twenty families. The Punan do not form a discrete ethnic group. "Punan" is a generic term for all groups of hunters and gatherers in Borneo, while "Dayak" refers to groups of shifting cultivators. The Punan are no longer nomads, however seasonal mobility remains high. Individuals, with or without their families, can move temporarily or definitively from one settlement to another, or migrate to East Malaysia for a month at a time or permanently. Whole villages can move at once, generally after an outbreak of a deadly epidemic disease. Many small villages were resettled often involuntarily by the government in the 1970s and regrouped with others in downstream locations, closer to towns and services (Kaskija 2002; Sellato 2001).

Historical factors resulted in strong differences amongst subgroups of Punan in terms of dialect, socioeconomic situation, subordination to overlords, and behavior towards outsiders. However, the Punan all originate from groups of hunter-gatherers who arguably only began upland swidden rice cultivation between the end of the nineteenth century and the middle of the twentieth. The Dutch pacification of East Borneo since the late nineteenth century put an end to an ancient and active headhunting tradition among Borneo peoples. This process opened the interior of the island to traders. With increased exchanges and less dependence on their Dayak neighbors as trading intermediaries, the Punan became more involved in the trade of forest products such as resins, gums, rattan, bezoar stones, and eaglewood. The decision to settle down implied the adoption of shifting cultivation, requiring families to stay in the vicinity of their upland rice swiddens during the rice-cropping season and even longer where rice is intercropped with cassava (Levang *et al.* 2002). The Punan's new lifestyle was influenced by that of their Dayak neighbors, including cultural traits such as the adoption of a more stratified social organization divided into aristocratic families, free men and bondsmen, the payment of bride price, and the capitalization of prestige goods (Chinese jars, copper gongs, and gold jewelry).

However, no group of Punan has ever been isolated, distinct, outside, and apart from the general pattern of Borneo's traditional life. As stressed by Hoffman (1984), the exchanges between the coast and the interior of Borneo during the last millennium had considerable effects on the mobility of every inland tribe. Large-scale socio-cultural changes have taken place over time, and the flow of people between different ethnic units and subsistence strategies has been considerable (King 1976).

From the mid 1970s to the end of the 1980s, the majority of the Punan of the Tubu River were resettled closer to town in large settlements like Respen Sembwak and Paking or in the mid-stream of the Malinau River (Kwala Ran, Long Loreh, Seturan). Others remained upstream but merged into larger villages such as Long Pada and Long Titi (Figure 7.10). In the space of two decades, nearly 10,000 households were pushed to resettle throughout Eastern Kalimantan (King 1993). Currently, 5,000 Punan live in the new district of Malinau where they represent less than 20 percent of the total population of the district, and nearly 1,000 live far upstream (Sitorus *et al.* 2004).

Although they share common cultural and technological traits, the Punan groups of the Malinau district are not related linguistically to their neighbors and do not share common origins (Kaskija 1999, 2002). The diversity of inter-ethnic relationship has shaped the local economic variations observable today in this area amongst the same population. The Punan of the Tubu have often intermarried with non-stratified groups of rice cultivators. Most of these groups have arguably inhabited the Bahau valley since the seventeenth century (Schneeberger 1979; Sellato 1995), from whence they entered the Tubu River. Such intermarriage and assimilation is rarely observed with highly stratified groups such as the Kayan and the Kenyah.

Kaskija (2002) argues that the Punan groups may have occupied the upper Tubu by 1800 and migration into the upper Malinau basin must have occurred by about 1850. According to him, four Punan Tubu subgroups can be distinguished in the Malinau area, depending on the type of collaboration established with Dayak groups: 1) Punan of the lower Tubu (now mainly located on the Mentarang) who were affiliated with the Abau and smaller groups such as the Milau, Mearu, and Merap; 2) Punan of the middle Tubu long associated with the Tebilun (Abai), an open non-stratified and fluid ethnic group from whom they adopted several features like the matrimonial compensation system; 3) Punan of the upper Tubu who had close links with the Merap; and 4) Punan of Malinau.

The distinction between Punan Malinau and Punan Tubu subgroups remains unclear and is complicated by frequent intermarriage. The differences could have resulted from the affiliation of Punan Malinau with the Merap, with whom they rarely intermarried. The Merap entered the Malinau-Tubu area from the west around 1880. For at least 150 years, the Merap have been the overlords of a large part of the Malinau and Tubu watersheds. This extremely stratified group did not allow much intermarriage, which explains

Figure 7.10. Current ethnic groups of the Tubu and Malinau river basins

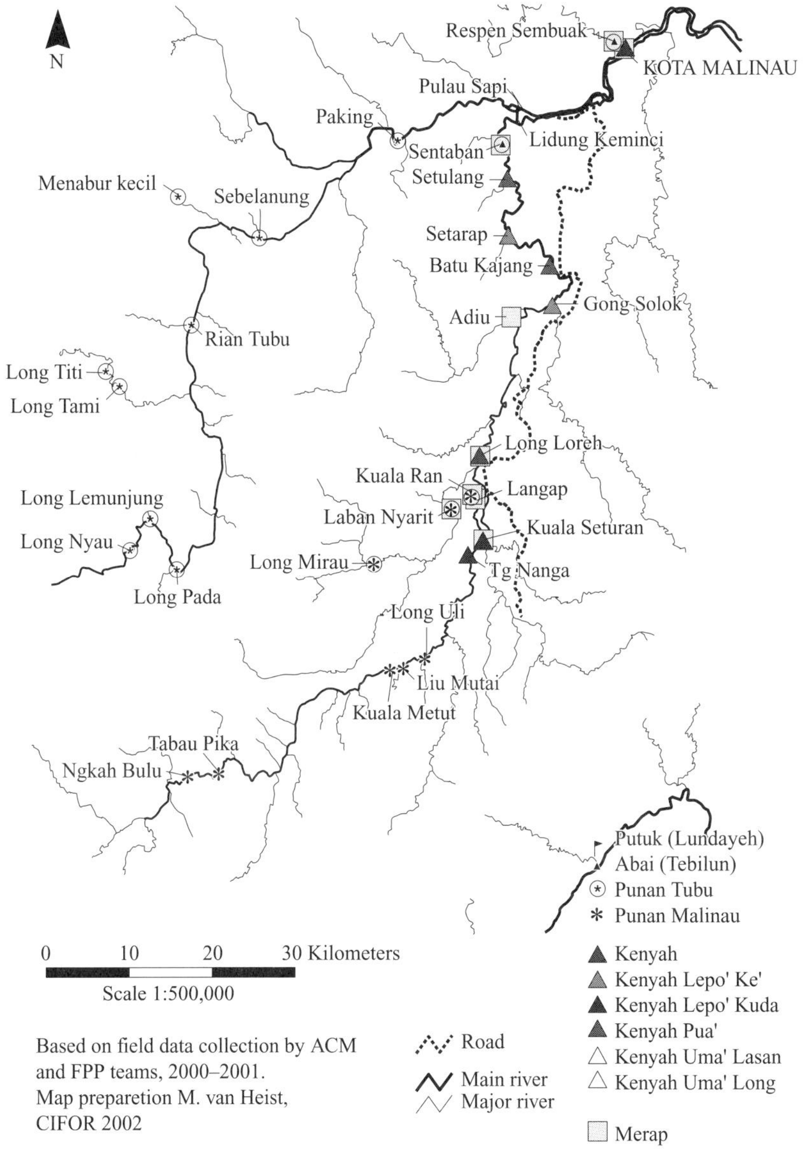

Source: van Heist (2002).

why the Punan of the upper Tubu and Malinau remained nomadic until a much later date (Sellato 1993). The Punan Tubu practiced rice cultivation earlier than the Punan Malinau who remained nomadic for a longer period

(Sellato 2001). The situation was reversed, however, in the last decade. The Punan of the uppermost Tubu are nowadays the only inhabitants of the Tubu River and appear more nomadic than the Punan Malinau who cohabitate on the river with diverse stratified groups of Putuk, Abai, Merap, and Kenyah (Figure 7.10).

Social and ecological features of contemporary Punan

Currently, the Punan are still seasonally mobile, exploiting a territory through repeated use of resource groves. This territory formally belongs to groups of swiddeners, to which the nomadic group was previously economically and socially affiliated. However, the nomads practice sustainable exploitation of the territory's subsistence resources including, for example, tending sago-palm groves. Punan groups, less territorially bound than farmers, often range far beyond the borders of the palm groves (Sellato, in press).

The Punan have a mixed subsistence economy that varies depending on locality. From upstream to the vicinity of Malinau, the Punan economy forms a "continuum of economic systems" (Dounias 2003; Kaskija 1999; Levang *et al.* 2002; Sellato 1994) displaying a variety of combinations of rice farming, horticulture, subsistence gathering, extractivism, fishing, hunting, and off-farm activities (for example transportation, guiding, working for mining and logging companies). The economy thus appears versatile with local intra-group variations in subsistence activities dependant upon political and historical causes of settlement (fusion/fission in alliances) and on remoteness from the main city of the district (upstream/downstream contrasts).

Patterns of commercial forest product collection and Punan migration

Trade and its shift from the coastal islands to the interior played a key role in Eastern Kalimantan's social and political changes between the seventeenth century and the turn of the twentieth (Sellato, 2001). Conflicts over traded forest products were far more common and violent than disputes over subsistence resources (Sellato, in press). Trade was driven by demand from distant markets, while middlemen linked the international traders to the interior of Kalimantan (Figure 7.11). The products that entered this trade included those used locally by inland peoples (such as rattan and damar) and those that had no local use (for example natural rubber, eaglewood, and birds' nests).

Downstream kingdoms were intermediaries in the international trade in forest products. These kingdoms were not territorially fixed; they moved according to trade opportunities and the establishment of new partnerships with leaders of the most isolated tribal groups upstream. Such partnerships were frequently bound by marriage and blood-brotherhood (Sellato 2001).

Traders pushed towards the interior, leading to the establishment of trade posts which would later become new petty kingdoms. Leaders of both coastal and inland trading groups actively acquired slaves in significant quantities to more systematically exploit the forest products within their territories. The late nineteenth century slave trade was primarily motivated by an increase in the demand for forest products. The Dutch finally abolished slavery in the 1910s. These and earlier pacification acts of the Dutch, however, pushed trade posts and middlemen further inland. Another result was that forest product collecting changed from an occasional to a year-round activity.

For the groups located farthest upstream, in places such as the Apo Kayan, the Bahau-Pujungan, or the Kerayan, involvement in forest product trade dates back to the end of the nineteenth century, corresponding with the intensification in the international demand for products like the latex of *Palaquium spp.* According to Sellato (2001), interior regions like Malinau remained relatively uninvolved in the forest product trade until the eighteenth century. Trade centers from Malinau began to emerge on the lower Tubu River in the 1920s and reached the remote villages only by the late 1950s (Figure 7.11). Collection and trade in forest products has remained dynamic until modern times. During the 1960s and 1970s *Banjir Kap* became a common practice that involved rural collectors. *Banjir* means "flood" in Indonesian, and *Banjir Kap* refers to the river-bound transport of logs during high water times. This recent phenomenon reflects the massive involvement of upriver people in the forest product trade. The banning of this logging technique, that is facing a revival today, led to the beginning of the latest round of collection of eaglewood from the interiors of Borneo.

Before the *Banjir Kap* period, forest product collection was rarely undertaken by Dayak agriculturalists. There were very few full-time collectors among stratified interior societies, and nomadic Punan bands dominated collecting. The leaders of swidden communities retained Punan groups to exploit forest products in their territories. Gathered products were brought to downstream markets on the occasion of the annual trade expedition (Sellato 2001). As a result of their engagement in forest product collection, Punan groups (sponsored by outsiders and receiving rice as payment) shifted from their customary nomadic life to the permanent search of forest products for trade. These groups also tracked down and killed outside professional collectors, or went on raids to steal forest products that were collected by other nomadic bands affiliated with different farming groups. The general situation was that the Punan were simply "guests" in formally bounded territories that were controlled by their stratified patrons.

The Punan satisfied their subsistence needs and obtained political and military protection from the farmers in exchange for trade items. However,

Figure 7.11. Progress of the NTFP extraction front in Eastern Kalimantan

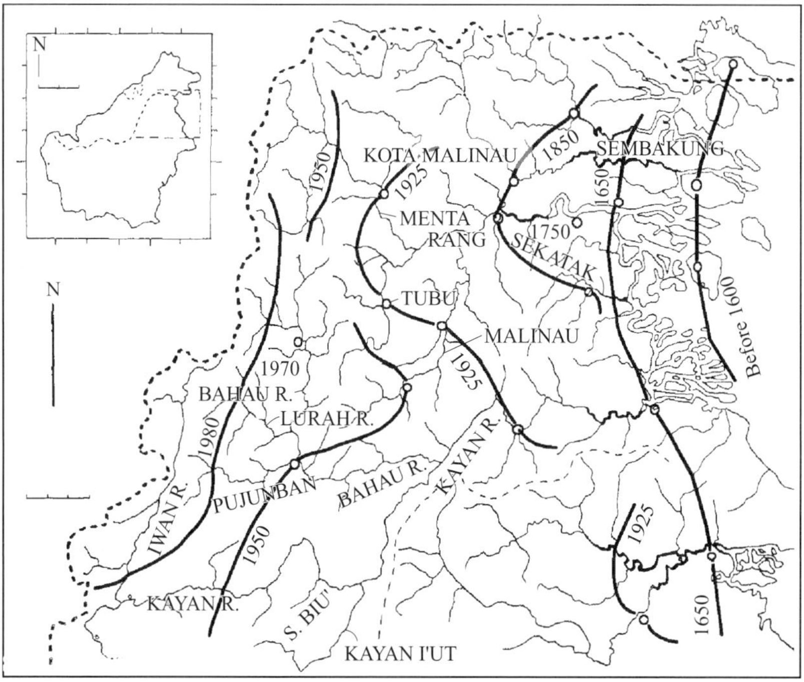

Reproduced from Sellato (2001).

the nature of interactions with socio-politically stratified neighbors varied from one community to another. The Punan associated with Tebilun, for instance, were known as very autonomous nuclear-family units that engaged in their own trading activities. By contrast, the Punan associated with Merap are described as more subservient towards their patrons. The Merap established numerous relations with Punan bands of the Tubu in the late nineteenth century. In addition to collecting, the Punan watched over highly valued birds' nest caves. Punan were also active in providing heads and slaves to their Merap leaders (Kaskija 2002).

Discussion

History of, history by, history for

The Baka and the Punan, as well as many other hunter-gatherer societies, underwent historical changes that were not unique to them alone, but were on the contrary part of wider regional historical trends. The historical reconstruction of discrete hunter-gatherer groups is a pointless endeavor due to the level of interaction that occurred between groups involved in the same economic activities and influenced by the same historical events (Klieman 1999). These connections have had a direct influence on contemporary mobility strategies. Both the Baka and the Punan now organize their social and spatial migration patterns on a seasonal basis from permanent villages. By settling in permanent villages, they shifted from a circulating to a radiating mobility strategy (Lieberman 1993). Migration now consists of frequent movements between villages, swidden huts, and forest camps. Paradoxically, it is assumed that the adoption of agriculture and its implied radiating mobility strategy has resulted in an increased and more complex spatial movement that requires in-depth planning (Binford 1980).

The present distribution of Pygmy groups in Africa is to a large extent dependent upon relations with local agriculturalists and outsiders—both influence the resource base of the hunter-gatherers presenting opportunities for and imposing constraints upon their activities. As a result, these relations are among the main factors influencing residence choices (Pedersen and Wæhle 1988). Despite the fact that there are substantial variations among Pygmy ethnic groups throughout the Congo Basin, almost no variation is observable in the composition of Baka local groups, even when relations with agriculturalists vary. Bahuchet (1993a) describes this as "the Pygmy phenomenon," referring to those people who borrowed their languages from their neighbors without changing their own culture. Presently, the Baka are not fragmented into different ethnic groups. On the contrary, similar village societies are grouped in patches inside a wide space uniformly occupied by the Baka. The Punan, however, present no ethnic uniformity. Their fragmented cultural traits and languages are testimony to local commercial alliances with particular stratified and non-stratified ethnic groups from whom they selectively adopted different cultural characteristics.

The ways of life of both Baka and Punan cannot be understood without also understanding their relations with neighboring farmers (Grinker 1994). In Central Africa, the complementary association between two specialized societies (Pygmy hunter-gatherers and neighboring farmers) is so efficient that it has persisted for a very long time, revealed through oral traditions, mythology, and the linguistic development of these groups (Bahuchet and Guillaume 1982). The Punan have a long history of trading forest products with other groups. This trade conditioned extensive Punan movements through the island. As a result of their involvement in trade, the Punan established political and economic associations with the aristocratic leaders of farming groups (Sellato 1994). They migrated from one watershed to another, either by following their partners or after disputes that led them to change alliances.

Hunter-gatherers arguably have unique histories (Bender and Morris 1988; Vansina 1986). However, a greater understanding of hunter-gatherer mobility must encompass knowledge of their links with farmers and the mobility patterns of the latter (largely influenced in Central Africa by the slave trade), as well as wider processes of capitalist penetration (such as further inland trade of forest products in Borneo). In the case of the Baka, slave trade was a dramatic event that initiated their currently fragmented distribution patterns two centuries ago, despite the fact that they were not directly affected by the slave trade. The slave trade drew them in via the close relationships they had with other ethnic groups that were raided by slave traders. The current distribution map of the Pygmies throughout southeastern Cameroon clearly bears the scars of this dramatic stage of history that occurred 250 years ago. In the case of the Punan, the successive episodes of forest product trading have continuously modified and built the complex distribution of the Punan groups until the late 1970s. Succeeding waves of collection and trade in forest products in the region have reshaped the distribution of the Punan of the Malinau district. As a result, current distribution patterns show little similarity with the ethnic scene that existed two centuries ago.

In, out, backwards, and forwards

The complexity of social and political systems has increased in recent years throughout tropical forests, as communities struggle to gain access to and control over economic resources. Competition and conflict have emerged on many levels, as the political economy and social structures change locally, regionally, nationally, and internationally. Within local settings, individuals of different social positions within communities compete for access to natural resources, as land—much more so than labor—increasingly becomes the primary constraint on production. Regional relations among ethnic communities have become particularly complicated with disputes over each other's access to lucrative resources, markets, and employment (Rupp 2002). Simultaneously, local communities are increasingly engaged in land-rights conflicts with local authorities and concessionaires.

In such contexts, the social ecology approach reveals its limitations, and historical analysis appears as a fundamental and complementary step to understanding the current complications of the inter-ethnic relations among contemporary communities and to help resolve local struggles for land-rights. Unlike the agriculturalist groups, Punan and Baka hunter-gatherers have not left many traces of their presence on the landscape. Nevertheless, they claim to be the original inhabitants of the regions they occupy. Undoubtedly, features of the history of forest exploitation by hunter-gatherers as well as by their farmer neighbors play an important role in the current management practices of subsistence resources. To make sense of any given tribal group's history, it is necessary to consider it as part of a wider regional history extending far beyond the boundaries of ethnic territories. Moving back and forth into the forest no longer adequately explains the migration patterns of forest dwellers. It is necessary to move backwards and forwards between past and present, along a scale that moves from the individual to the group and all the way to geopolitics.

Conclusions

The practical understanding of past and current relationships among environmental and human systems arguably requires a temporal and spatial perspective of landscape be applied at a regional level. The landscape retains the physical evidence of cognitive activities, notably the decision to maintain or modify particular environmental features or migration patterns. The historical ecology of landscape offers a comprehensive interdisciplinary framework to assess the temporality of complex interactions between nature and culture and to suggest directions of change.

As environmental degradation is undoubtedly the most pressing and potentially disastrous problem facing forest dwellers, consideration of their mobility at broader temporal and spatial scales seems appropriate in dissecting complex chains of mutual causation in human-environment relations. Changes in subsistence strategies, demography, or perception have through time resulted in both intentional and unintentional modification of the forests. Given the rapid deterioration of tropical forests, it is extremely important that we draw on our knowledge of historical events that influence the migration of hunter-gatherer societies, and subsequently affect their contemporary exploitative and managerial practices.

Acknowledgements

The authors would like to express their warm thanks to the organizers of the Symposium "In, Out, In, Out: Populations, Migrations, and the Social Ecology of Tropical Forests" for inviting us to Kyoto in December 2002. We eventually could not attend the symposium, but we are grateful that the editors offered us the possibility of submitting a paper to be published in this book. Special thanks to Lye Tuck-Po and Wil de Jong who did a wonderful job editing our work by transforming our "Franglish" into proper English, and our first draft into a publishable paper.

References

Alexandre, P. 1965. Protohistoire du groupe Beti-Bulu-Fang: Essai de synthèse provisoire. *Cahier d'Etudes Africaines*, 5(20): 503–560.

Alexandre, P. and J. Binet. 1958. *Le Groupe dit "Pahouin" (Fang, Bulu, Béti)*. Paris: Presses Universitaires de France.

Bahuchet, S. 1985. *Les Pygmées Aka et la Forêt Centrafricaine: Ethnologie Ecologique*. Paris: SELAF-CNRS.

Bahuchet, S. 1992. Spatial mobility and access to resources among African Pygmies. In M.J. Casimir and A. Rao eds., *Mobility and Territoriality: Social and Spatial Boundaries Among Foragers, Fishers, Pastoralists and Peripatetics*, 205–257. Oxford: Berg.

Bahuchet, S. 1993a. *Histoire d'une Civilisation Forestière, Tome 2: La Rencontre des Agriculteurs: Les Pygmées Parmi les Peuples d'Afrique Centrale*. Paris: Peeters-SELAF.

Bahuchet, S. 1993b. History of the inhabitants of the Central African Rain Forest: Perspectives from comparative linguistics. In C-M. Hladik, H. Pagezy, O.F. Linares, A. Hladik, A. Semple, and M. Hadley, eds. *Tropical Forests, People and Food: Biocultural Interactions and Applications to Development*, 37–54. Paris: Unesco-Parthenon.

Bahuchet, S. and H. Guillaume. 1982. Aka–farmer relations in the northwest Congo Basin. In E. Leacock and R.B. Lee, eds., *Politics and History in Band Societies*, 189–211. Cambridge: Cambridge University Press.

Balée, W., ed. 1998. *Advances in Historical Ecology*. New York: Columbia University Press.

Barnard, A. 1983. Contemporary hunter-gatherers: Current theoretical issues in ecology and social organization. *Annual Review of Anthropology*, 12: 193–214.

Bender, B. and B. Morris. 1988. Twenty years of history, evolution and social change in gatherer-hunter studies. In T. Ingold, D. Riches, and J. Woodburn, eds., *Hunters and Gatherers, vol. 1: History, Evolution, and Social Change*, 4–14. Oxford: Berg.

Biesbrouck, K., S. Elders, and G. Rossel, eds. 1999. *Central African Hunter-Gatherers in a Multidisciplinary Perspective: Challenging Elusiveness*. Leiden: Research School CNWS.

Binford, L.R. 1980. Willow smoke and dogs' tails: Hunter-gatherer settlement systems and archaeological site formation. *Journal of Anthropological Archaeology*, 1(1): 4–20.

Boomgaard, P. 1997. Introducing environmental histories of Indonesia. In P. Boomgaard, F. Colombijn, and D. Henley, eds., *Paper Landscapes: Explorations in the Environmental History of Indonesia*, 1–26. Leiden: KITLV Press.

Boomgaard, P. 1998. The VOC trade in forest products in the seventeenth century. In R.H. Grove, V. Damodaran, and S. Sangwan, eds., *Nature and the Orient: The Environmental History of South and Southeast Asia*, 375–395. Delhi: Oxford University Press.

Burnham, P., E. Copet-Rougier, and P. Noss. 1986. Gbaya et Mkako: Contribution ethnolinguistiques à l'histoire de l'est Cameroun. *Paideuma*, 32: 87–128.

Burnham, P., and R.F. Ellen, eds. 1979. *Social and Ecological Systems*. London: Academic Press.

Butzer, K.W. 1982. *Archaeology as Human Ecology: Method and Theory for a Contextual Approach*. Cambridge: Cambridge University Press.

Casimir, M.J. and A. Rao, eds. 1992. *Mobility and Territoriality: Social and Spatial Boundaries Among Foragers, Fishers, Pastoralists and Peripatetics*. Oxford: Berg.

Copet, E. 1981. Les Kaka. In E. Tardits, ed., *Contribution de la Recherche Ethnologique à l'Histoire des Civilisations du Cameroun*, 511–516. Paris: Editions du CNRS.

Copet-Rougier, E. 1987. Du Clan à la Chefferie dans l'est du Cameroun. *Africa*, 57(3): 345–363.

Coquery-Vidrovitch, C. 1972. *Le Congo au Temps des Grandes Compagnies Concessionnaires*. Paris-La Haye: Mouton.

Crumley, C.L., ed. 1993. *Historical Ecology: Cultural Knowledge and Changing Landscapes*. Santa Fe: School of American Research Press.

Dhellemmes, I. 1977–1982. *Rapports sur la Situation Démographique des Pygmées Baka (8 Rapports Relatifs aux Divers Arrondissements)*. Sangmelima: Evéché de Sangmelima.

Dieu, M. and P. Renaud, eds. 1983. *Inventaire Préliminaire—Cameroun*. Paris: Projet ALAC (ACCT-CERDOTOLA).

Dounias, E. 2001. Les Tarières à Ignames Sauvages des Pygmées Aka et Baka d'Afrique Centrale. *Techniques et Culture*, 37: 127–154.

Dounias, E. 2003. *Wild Boar, Seasonality, and Diet: Coping with Changes in Food Supply Among the Punan Tubu of Eastern Kalimantan*. Paper presented at: XV ICAES Congress, Commission on the Anthropology of Food (ICAF), "Anthropology, Nutrition, and Wildlife Conservation", July 5–12, 2003, Florence.

Eghenter, C. 1999. Migrants' practical reasonings: The social, political, and environmental determinants of long-distance migrations among the Kayan and the Kenyah of the interior of Borneo. *Journal of Social Issues in Southeast Asia*, 14(1): 1–33.

Ellen, R.F. 1989. *Environment, Subsistence and System: The Ecology of Small-Scale Social Formations*. Cambridge: Cambridge University Press.

Ellen, R.F and K. Fukui, eds. 1996. *Redefining Nature: Ecology, Culture and Domestication*. Oxford: Berg.

Geschiere, P.H. 1981. Remarques sur l'histoire des Maka. In C. Tardits, ed., *Contribution de la Recherche Ethnologique à l'Histoire des Civilisations du Cameroun*, 517-531. Paris: Editions du CNRS.

Gremillion, K.J., ed. 1997. *People, Plants, and Landscapes: Studies in Paleoethnobotany*. Tuscaloosa: University of Alabama Press.

Grinker, R.R. 1994. *Houses in the Rainforest: Ethnicity and Inequality Among Farmers and Foragers in Central Africa*. Berkeley: University of California Press.

Guillaume, H. 2000. *Du Miel au Café, de l'Ivoire à l'Acajou: La Colonisation de l'Interfluve Sangha-Oubangui et l'evolution des Rapports Entre Chasseurs-Collecteurs Pygmés Aka et Agriculteurs (Centrafrique, Congo) 1880–1980*. Paris: Peeters-SELAF.

Harms, R. 1981. *River of Wealth, River of Sorrow: The Central Zaïre Basin in the Era of the Slave and Ivory Trade, 1500–1891*. New Haven: Yale University Press.

Hitchcock, R.K. 1982. *The Ethnoarchaeology of Sedentism: Mobility Strategies and Site Structure among Foraging and Food Producing Populations in the Eastern Kalahari Desert, Botswana*. PhD dissertation, University of New Mexico, Albuquerque.

Hoffman, K.L. 1984. Punan foragers in the trading networks of Southeast Asia. In C. Schrire, ed., *Past and Present in Hunter-Gatherer Studies*, 123–149. Orlando: Academic Press.

Hutterer, K.L. 1983. The natural and cultural history of Southeast Asian agriculture: Ecological and evolutionary considerations. *Anthropos*, 78: 169–212.

Ichikawa, M. 1999. Interactive process of man and nature in the Ituri forest of the Democratic Republic of Congo: An approach from historical ecology. In K. Biesbrouck, S. Elders, and G. Rossel, eds., *Central African Hunter-Gatherers in a Multidisciplinary Perspective: Challenging Elusiveness*, 141–152. Leiden: Research School CNWS.

Ingold, T., D. Riches, and J. Woodburn, eds. 1988. *Hunters and Gatherers, vol. 1: History, Evolution, and Social Change.* Oxford: Berg.

Kaskija, L. 1999. The Punan of Borneo: Cultural fluidity and persistency in a forest people. In A. Hornborg and M. Kurkiala, eds., *Voices of the Land: Identity and Ecology in the Margins.* Lund: Lund University Press.

Kaskija, L. 2002. *Claiming the Forest. Punan Local Histories and Recent Developments in Bulungan, East Kalimantan.* Bogor: CIFOR.

Kent, S., ed. 1996. *Cultural Diversity Among Twentieth Century Foragers: An African Perspective.* Cambridge: Cambridge University Press.

Kent, S. and H. Vierich. 1989. The myth of ecological determinism: Anticipated mobility and site spatial organization. In S. Kent, ed., *Farmers as Hunters: The Implications of Sedentism*, 96–130. Cambridge: Cambridge University Press.

King, V.T. 1976. Migration, warfare, and culture contact in Borneo: A critique of ecological analysis. *Oceania*, 46: 306–327.

King, V.T. 1993. *The Peoples of Borneo.* Oxford: Blackwell.

Klieman, K. 1999. Hunter-gatherer participation in rainforest trade systems: A comparative history of forest vs. ecotone societies in Gabon and Congo, c. 1000–1800 A.D. In K. Biesbrouck, S. Elders, and G. Rossel, eds., *Central African Hunter-Gatherers in a Multidisciplinary Perspective: Challenging Elusiveness*, 89–104. Leiden: Research School CNWS.

Laburthe-Tolra, P. 1981. *Les Seigneurs de la Forêt.* Paris: Publications de la Sorbonne.

Laden, G.T. 1992. *Ethnoarchaeology and Land Use Ecology of the Efe (Pygmies) of the Ituri Rain Forest, Zaire: A Behavioral Ecological Study of Land Use Patterns and Foraging Behavior.* PhD dissertation, Harvard University, Cambridge.

Leacock, E. and R.B. Lee, eds. 1982. *Politics and History in Band Societies.* Cambridge: Cambridge University Press.

Leclerc, C. 2001. *En Bordure de Route: Espace Social, Dynamisme et Relation à l'environnement chez les Pygmées Baka du Sud-Est Cameroun.* Thèse de doctorat, Université de Paris X-Nanterre, Paris.

Lee, R.B. and I. DeVore, eds. 1968. *Man the Hunter.* Chicago: Aldine.

Levang, P., and the FPP-Bulungan Team. 2002. People's dependencies on forests. In *Forest, Science and Sustainability: The Bulungan Model Forest Technical Report, Phase 1, 1997–2001*: ITTO Project PD 12/97, 109–130. Bogor: CIFOR.

Levang, P., E. Dounias, and S. Sitorus. In press. *Out of Forest, Out of Poverty? Forests, Trees, and Livelihoods*, 15(2).

Lieberman, D.E. 1993. The rise and fall of seasonal mobility among hunter-gatherers: The case of the southern Levant. *Current Anthropology*, 34(5): 599–631.

Martin, G. 1948. *Histoire de l'esclavage dans les Colonies Françaises*. Paris: Presses Universitaires de France.

Mohammadou, E. 1978. *Les Royaumes Foulbe du Plateau de l'Adamaoua au XIXe Siecle; Tibati, Tigne, Banyo, Ngaoundere*. Tokyo: Institute for the Study of Languages and Cultures of Asia and Africa.

Mohammadou, E. 1991. *Traditions Historiques des Peuples du Cameroun Central, vol. 2: NiZoo, Voute et Kondja*. Tokyo: Institute for the Study of Language and Culture of Asia and Africa.

Monino, Y. 1995. *Le Proto-Gbaya: Essai de Linguistique Comparative sur Vingt-et-une Langues d'Afrique Centrale*. Paris: Peeters-SELAF.

Pedersen, J. and E. Wæhle. 1988. The complexities of residential organization among the Efe (Mbuti) and the Bamgombi (Baka): A critical view of the notion of flux in hunter-gatherer societies. In T. Ingold, D. Riches, and J. Woodburn, eds., *Hunters and Gatherers, vol. 1: History, Evolution, and Social Change*, 75–90. Oxford: Berg.

Rupp, S.K. 2002. *I, You, We, They: Forests of Identity in Southeastern Cameroon*. PhD dissertation, Yale University, New Haven.

Schneeberger, W.F. 1979. *Contributions to the Ethnography of Central Northeast Borneo (Parts of Kalimantan, Sarawak and Sabah)*. Berne: University of Berne.

Schrire, C., ed. 1984. *Past and Present in Hunter-Gatherer Societies*. New York: Academic Press.

Sellato, B. 1993. The Punan question and the reconstruction of Borneo's culture history. In V.H. Sutlive, Jr., ed., *Change and Development in Borneo*, 47–81. Williamsburg: Borneo Research Council.

Sellato, B. 1994. *Nomads of the Borneo Rainforest: The Economics, Politics, and Ideology of Settling Down*. Honolulu: University of Hawaii Press.

Sellato, B. 1995. *The Ngorek: A Survey of Lithic and Megalithic Traditions in the Bahau Area, East Kalimantan, and an Interdisciplinary Sketch of Regional History*. Jakarta: Culture and Conservation, World Wide Fund for Nature Indonesia Programme.

Sellato, B. 2001. *Forest, Resources and People in Bulungan: Elements for a History of Settlement, Trade and Social Dynamics in Borneo, 1880–2000*. Bogor: CIFOR.

Sellato, B. In press. Forests for food, forests for trade, between sustainability and extractivism: The economic pragmatism of traditional peoples and the trade history of northern East Kalimantan. In R.L. Wadley, ed., *Histories of the Borneo Environment: Economic, Political, and Social Dimensions of Transformation*. Leiden: KITLV Press.

Sitorus, S., P. Levang, E. Dounias, D. Mamoung, and D. Abot. 2004. *Potret Punan Kalimantan Timur: Sensus Punan 2002–2003*. Bogor: CIFOR.

Soret, M. 1978. *Histoire du Congo*. Paris: Berger-Levrault.

Tardits, C. 1980. *Le Royaume Bamoum*. Paris: Armand Colin.

Tardits, C., ed. 1981. *Contribution de la Recherche Ethnologique à l'histoire des Civilisations du Cameroun*. Paris: Editions du CNRS.

Vansina, J. 1986. Do Pygmies have a history? *Sprache und Geschichte in Afrika*, 7(1): 431–445.

Vansina, J. 1990. *Paths in the Rainforests: Toward a History of Political Tradition in Equatorial Africa*. London: James Currey.

8
An Overview of Migration and Deforestation in Southeast Asia: 1880 to 2002

Lesley Potter

Introduction

Southeast Asia provides an interesting laboratory for studying the human dimensions of migration and deforestation. The region originally possessed an enormous forest resource, extending from the dipterocarp-dominant forests of the equatorial zone, through mixed deciduous stands, sometimes including teak growing under monsoonal conditions. Littoral and swamp forests of mangrove and associated species were extensive in coastal regions. As populations increased, forests slowly retreated. Lands were converted partly to food crop production, although export cash crops have triggered deforestation in both historical and modern periods. While pepper and spice cultivation induced early forest clearance, the Dutch Cultivation System, with its forced deliveries of coffee, transformed the mountain forests of Java (Hefner 1990) and parts of upland Sumatra. From about the 1880s, the rate of conversion grew; deltaic forests in Burma became rice bowls and plantation tobacco reached its heyday in eastern Sumatra. In the early twentieth century, rubber began to replace the dipterocarps in foothill areas of the Federated Malay States. Such transformations were accompanied by large-scale migration as labor was moved to the cash crop frontiers from within and beyond national boundaries. Areas with already dense populations were targeted for sponsored movements, while spontaneous outflows were encouraged through improved accessibility.

Modern migration and deforestation rates increased from the 1970s through the 1990s, following the commercialization of the dipterocarp forests and the political upheavals during and after the Indochina War. Population growth rates reached their maximum during the 1970s and have since been declining, although total numbers and densities have continued to rise, including densities in the forested zones. Major changes are now

Table 8.1. Major population and land-use/land-cover statistics, Southeast Asia[a]

Country	Population (million) 1980	1990	2000	% land under permanent crop (1999)	% land natural forest (2000) (FAO)	% land forest plantation (2000) (FAO)
Mainland SE Asia						
Cambodia	6.5	8.6	13.1	21.6	52.4	0.5
LaoPDR	3.2	4.2	5.3	4.1	54.2	0.2
Myanmar	33.6	40.8	49.0	15.4	51.1	1.2
Thailand	47.0	56.3	61.8	35.2	19.2	9.6
Vietnam	53.7	66.2	77.7	22.6	24.9	5.2
Sub-total	144.0	176.1	206.9			
Insular SE Asia						
Brunei	0.19	0.25	0.34	1.3	83.3	0.6
Indonesia	147.5	179.4	210.5	17.1	52.5	5.4
Malaysia	13.9	18.1	23.3	23.1	53.4	5.3
Philippines	48.3	61.5	76.5	33.7	16.9	2.5
Singapore	2.4	3.1	4.0	2.9	3.3	-
Sub-total	212.3	262.4	314.6			
Total	356.3	438.5	521.5			

Source: ESCAP (2001), FAO (2000). Separate figures for East Timor were not available from ESCAP.

a: Table 8.1 summarizes major parameters of population, land-use and land-cover, with forest statistics drawn from the most recent FAO study (FAO 2000). Unfortunately these figures are not comparable over time, as the FAO changed its method of calculation in 2000, reverting to a reliance on national estimates rather than mathematical modeling. This led in retrospect to higher forest cover figures for 1990, replacing earlier published data. The FAO is now presenting "net" figures for forest cover, offsetting forest loss with areas reforested (usually in single species plantations). Observers argue that these by no means represent a suitable substitute for natural forests (Matthews 2001). Definitions of forest area are notoriously difficult to compare, and many of the developing country data used were either old or incomplete. However, precise forest cover figures are not critical for this paper, in which interest lies more with the migrants and the processes at work on the forest frontiers.

occurring as agricultural frontiers close in countries such as Thailand, the Philippines, and Vietnam, and reforestation policies have begun to restrict agricultural expansion. The forests in countries such as Indonesia continue to offer resources, but on a shortened time-scale.

In this chapter I offer an overview of processes of migration and deforestation in Southeast Asia from the late nineteenth century to the present. I demonstrate a wide range of interactions between economic, political, and demographic factors that have influenced migration into and out of Southeast Asian forests. In the second section of the chapter I illustrate the way in which migration can be built into three explanatory frameworks for understanding tropical deforestation. In addition, I propose a fourth framework. In sections three, four, and five I analyze the complex interplay between factors involved in forest exploitation and migrational patterns, using historical and contemporary case studies. In section six I explain these cases, employing the frameworks outlined in section two. In section seven I present the conclusions.

Theoretical approaches to deforestation

Wunder (2002) recognizes three schools of thought concerning the causes of tropical deforestation, each of which attributes a different role to migration:

The neo-classical school

The neo-classical school sees profit maximization as the incentive underlying deforestation, with a mobile labor-force seeking the best available options. Clearing land for agriculture is more profitable than leaving it forested. High prices for timber or non-wood forest products also encourage in-migration, ultimately leading to degradation and subsequent clearing. High commodity prices encourage the cash-crop smallholder to clear more forest and pioneers to migrate to the frontier, while fluctuating market demands for agricultural commodities may promote itinerant migration behavior. Colombijn (1997: 324) demonstrated this phenomenon in the case of Chinese gambier producers in the Riau Archipelago from 1830 to 1870. During periods of rising prices forest conversion expanded rapidly, while an economic slump led to the temporary abandonment of gardens: "the cultivators moved like migratory birds, coming or going with the rise or fall of the price in the main market." Modern smallholder coffee production in Indonesia and Vietnam showed initial rapid expansion into natural forests. After returns dropped below production costs, some producers totally abandoned coffee and withdrew from the "frontier", most farmers, however, tried additional or alternative crops, collected forest products, or found temporary work elsewhere. Some invaded new forest areas, seeking higher coffee yields to offset the low prices (source: fieldwork, 2001).

According to de Koninck and Dery (1997), agricultural expansion associated with in-migration is the central factor driving deforestation in Southeast Asia. While this was true in the past, agricultural expansion may have reached its limits in a number of countries, with the pace of migration into rural areas now much reduced. This is apparent in both Thailand and the Philippines, where the proportion of remaining natural forest is quite low and policies are directed more toward reforestation and agroforestry. Out-migration to urban areas has become the norm (Missingham 1996; Delang 2002). In Peninsular Malaysia, agricultural expansion first replaced the forest with rubber and then oil-palm. Malay families moved to tree crop-based land settlement schemes during the 1970s, but their children prefer urban occupations. Many smallholder schemes have reverted to plantations, using imported Indonesian labor. Although agricultural expansion continues in Eastern Malaysia, Peninsular firms seeking forested land for oil-palm find it offshore—in Indonesia. Despite the periodic expulsion of illegal immigrants by Malaysia, a symbiotic relationship of labor and land exchange continues between the two states.

The neo-Malthusian school

The neo-Malthusian school views population as the driving force behind resource depletion. In some cases this school simplistically equates higher levels of population growth or density with forest destruction. It is true that at the regional level, population density and forested land are negatively correlated. Higher population levels may be caused by migration, either sponsored or spontaneous, and/or by natural growth. Based solely on estimates of population growth, Fraser (1998) predicted an annual deforestation rate of 900,000 to 1,000,000 ha for Indonesia, which is only half the actual rate (Barber *et al.* 2002). Governmental, economic, socio-political (or cultural), and technological factors modify the link between population and deforestation (Geist and Lambin 2001: 96). Direct governmental impacts on population and migration include road construction, agricultural subsidies and tax incentives, gasoline prices, and macro-economic policies (Kaimowitz and Angelsen 1998: 95). Kummer (1992: 65) sees deforestation in the Philippines as a two-stage process, with logging companies first creating accessibility to upland forest areas, followed by migrants carving out farms along the logging roads.

Another branch of the population-deforestation school correlates the impoverishment of local people, whose survival depends on plundering forests, with population growth. Studies have suggested that forest clearing was often beyond the resources of the poor, unless they managed to accumulate capital—specifically labor and time (Wibowo and Byron 1999: 469; Angelsen and Resosudarmo 1999: 17). It has been argued that in frontier situations, often associated with high in-migration rates, forest clearing is more likely to be "capital-driven", whereas in other highly populated areas poverty can in fact induce deforestation (Geist and Lambin 2001: 72–75). Rudel and Roper (1997: 53) distinguish "frontier theory", where large forests still exist and entrepreneurs, companies, and small farmers are the main agents of destruction, from "immiseration theory", where small areas of forest are attacked by expanding peasant populations with no alternatives. These two theories, however, presume a lack of alternatives, such as labor migration, "deagrarianisation"[1] (Rigg and Nattapoolwat 2001), or cash cropping.

Political ecology

Political ecology explanations are concerned with the role of government and the nature of governmental policy regarding forests, the functioning of institutions charged with implementing forest policy, the inequitable distribution of benefits from the exploitation of forest resources, and the territorial power relationships on the forest frontier. Political ecology analysis connects deforestation to political processes, looking at, for example, government policies that grant forest access to powerful capitalists engaged in rent-seeking or favor migrants over indigenous inhabitants. Such policies are the direct result of skewed power relations and may potentially provoke confrontation, violence, and arson, as well as more symbolic forms of protest. For example, the change in the political climate in post-Suharto Indonesia, from coercion and control to "reformation" and lax rule enforcement, has led to an explosion of "illegal logging" by previously dispossessed villagers and others seizing the opportunity. Political opposition may be expressed in various ways. Thai tree ordination rituals are designed with complex motives: to protect forests from commercial logging, to demonstrate local capacity to manage forest resources, to counter government territorialization, and to contest stereotypes of forest-dwellers as forest destroyers (Isager and Ivarsson 2002: 396). More direct and violent confrontations between rival contenders for forest resources are not uncommon (Hirsch 1988; EIA and Telapak 1999).

Social ecology

I suggest a fourth dimension to the previous list; that of *social ecology*. This approach examines the relationship between human populations and their environments. Where migrants are concerned, it refers to the perceptions and reactions of people exposed to new and different natural environments, such as forest margins, but also new social environments, in which they must relate to and share resources with indigenous groups. Both internal and international migrations are widespread in Southeast Asia, and have operated throughout the period under review. During colonial times, there was a massive transfer of labor from outside the region to support agricultural plantation enterprises, bringing Indians to Malaya and Chinese to East Sumatra. Even where the new migrants were from the same colonial territory, there were usually large differences between their home environments and those of the forested frontier. The historical literature is replete with stories of the difficulties of settlers in "conquering" the forest, evident in the following examples from the Burma delta and Vietnam's rubber belt. In the modern period, transmigrants from Java have had to learn how to handle the forests of Kalimantan, more successfully accomplished where local practices were adopted. Relations with indigenous groups were often tense and at times resulted in fatalities. Everywhere many migrants failed, while others, although planning permanent residence, were forced to retreat.

The next three sections present case studies of particular forest-ward migrations, discussing economic, political, and demographic factors, as well as the overall social ecology of the area.

Late nineteenth and early twentieth century frontier migrations

Settling the Irrawaddy delta

Adas has noted the rapid transformation of Lower Burma (Figure 8.1) in the second half of the nineteenth century "from an underdeveloped and sparsely populated backwater into the world's leading rice-exporting area" (1974: 4). This transformation was partly achieved through a massive wave of migration from Upper to Lower Burma after 1885, with 90 percent of new settlers originating from the north.[2] As the alluvial Upper Delta became filled, there was a shift to the Lower Delta—a mass of islands fringed with thickly vegetated mangrove swamps, clothed with *kanazo* forest (*Heritiera fomes*) and interspersed with open stretches of elephant grass (*Saccharum spontaneum*) (Adas 1974: 10). Migrants found an inhospitable environment of jungle and swamp, harboring tigers, wild elephants, and poisonous snakes. They fortified their houses to ward off dangerous animals, but their fields lay open to invasions of monkeys, rats, and birds. Malaria influenced settlement patterns, as anophelene mosquitoes predominated in new forest clearings (Adas 1974: 25). Most migrants initially "squatted" on a piece of jungle or scrub which they gradually cleared and cultivated, later securing tenure by paying land tax. *Kanazo* trees, growing up to 50 m tall, had to be felled and burnt before cultivation could begin. After the first year's attack on the forest, settlers planted sugar cane among the roots for one to two years. It was then possible to plough and plant rice, but the land did

Figure 8.1. Locations mentioned in the chapter

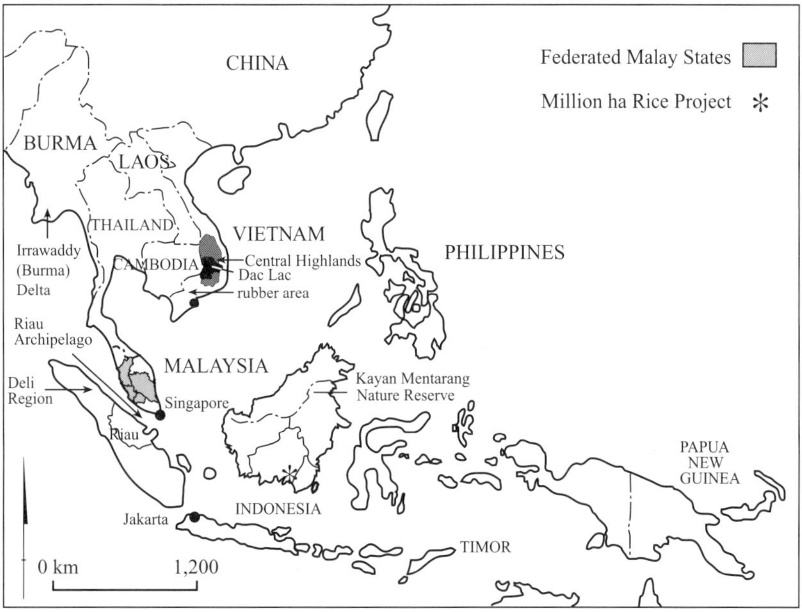

not reach full productivity for another four to five years. What prompted them to undertake such an enterprise? The initial impetus was the unsettled state of Upper Burma in 1885, with food shortages compounded by drought. Many migrants began as seasonal laborers, moving in and out of the region until they made the decision to stay. Initially, almost all were male; bringing families only after the land became productive. However, an assured export market for rice, already achieved by the 1870s, gave promise of financial gain. The government assisted with drainage and transport, providing plough animals and improved tools. While *kanazo* trees had supplied timber for construction, firewood, and charcoal, forest resources counted for little in the face of the world market for rice. It was market demand that sealed the fate of the forest (Adas 1983: 104).

Migrations and plantation crops

The plantation industries which were established in East Sumatra from the 1870s—first tobacco, later rubber, tea, and oil-palm—created an artificial landscape into which workers were imported, initially from China, then mainly from Java as "contract coolies". Although some worked in preliminary clearings, their basic social ecology was that of a life restricted in large part to plantation activities until eventual repatriation. Despite such restrictions, some Javanese, Banjarese, and others remained after their contracts ended, their small agricultural settlements extending into local forests. The district's population increased from 100 thousand in 1880 to 1.5 million in 1930 (Stoler 1985).

The first decade of the twentieth century saw the establishment of rubber plantations in the Federated Malay States, especially Perak, Selangor, and Negri Sembilan. The plantations, laid out in the forested western foothills of the Main Range within reach of the transport system servicing tin mines, led to considerable deforestation. It was estimated that 50 million tons of timber was burnt by the woodcutting gangs in two decades of clearing (Foxworthy 1921). By 1911, "the whole face of the country, as seen from our main roads, has been transformed from virgin forest to the monotony of miles of rubber trees" (Government of Perak 1911: paragraph 17). The rubber estates also imported contract labor, in this case mainly from India. This large labor-force in the foothills (250,000 workers) provided an ideal source of infection for the malaria vector which proliferated in the ravines: the death toll among plantation laborers was very high (Wylie 1993).

Rubber plantations were also established in Cochinchina (Vietnam) in the 1920s, primarily in the tall dipterocarp forest north of Saigon. The new labor-force, recruited from Tonkin in the north, had to first clear the forest before planting could begin. Tran (1985: 25) provides a graphic account of these clearing activities:

> First of all, we workers had to fell the trees, clear out the underbrush, dry everything in the sun, and then burn it. It was extremely hazardous work, felling those giant hardwoods and oil trees, which had branches reaching out who knows how far above. Each time a tree was felled and came crashing down, the rushing sound of the branches lashing the air was terrifying to hear.

Workers frequently died or were injured while tree-cutting.[3] As in Malaya, malaria was a constant scourge.

The role of the state and the forest frontier: Selected migrations post-1970

Despite the earlier movements discussed above, large areas of forest remained relatively intact after World War II. A "new great age of clearance" and forest settlement was about to begin (Brookfield 1988). This new forest invasion accompanied the first dipterocarp logging boom in Insular Southeast Asia, while on the Mainland it took place in the aftermath of the Indochina War. Much of this was government initiated, but the role of the state varied enormously, from a very strong and controlling presence in the Malaysian FELDA schemes, through various degrees of spontaneous migration. Spontaneous migration has increased in importance and gradually replaced state-assisted schemes in Indonesia and Vietnam.

Transmigration in Indonesia

The more recent period of government-sponsored and spontaneous migration in Indonesia is partly an intensification of preexisting movement patterns. The Dutch colonial government's settlements from Java to the *tanah sabrang* ("the land beyond") started early in the twentieth century (Pelzer 1948). Transmigration became a leading policy of the Suharto regime, aimed at population redistribution and regional development, but also at establishing Javanese cultural hegemony and border security. Table 8.2 indicates a movement of almost 7 million people from 1969 to the end of 1999. Since its inception, the focus of the program has shifted in a number of ways: difficult tidal swamps are no longer the preferred location; failed food crops have been replaced with tree crop-based schemes; and governmental involvement has decreased.[4] The transmigration program itself was officially terminated in 2001, but lingers in the form of "local transmigration" and refugee resettlement. East Timorese former refugees still in camps in West Timor, for instance, have been offered the option of transmigration to the nearby island of Sumba, or to Central Kalimantan (*Jakarta Post* 16.9.02).

Hidayati (1994), researching transmigrants in food-crop schemes in Kalimantan, studied the interactions between the newcomers and two kinds of environments: dry uplands and tidal swamps. After their first three years the upland settlers were forced by declining yields to adopt swiddening techniques. Instead of using the long fallow swidden which was

Table 8.2. Indonesia: Estimated number of transmigrants, 1969–1999

Period	1969–74	1974–79	1979–84	1984–89	1989–94	1994–99	1999 (to November)
Targeted families	38,700	250,000	500,000	750,000	550,000	600,000	16,235
Families that moved	36,483	118,000	535,000	230,000	247,000	300,000	4,409
Total people	174,000	544,000	2,469,560	1,061,680	1,136,200	1,500,000	22,000

Source: Adhiati Adriana and Bobsien (2001); Mubyarto (2000).

still common in the district, they chose short fallows and fairly continuous cultivation, eliminating returning tree cover and effectively converting the land to *Imperata* grass, necessitating further clearing. In the tidal swamps the transmigrants tried to use techniques suitable for conditions in Java such as deep tillage. After they too were faced with declining yields and increasing acidification, they turned to local Banjarese for help and began to use traditional shallow cultivation methods to suit the new conditions (Hidayati 1994). Oekan (1996), studying a group on a tidal swamp settlement, concluded that these people had become "opportunists", looking for short-term gains wherever they could be found. Some made a living by clearing the surrounding *Melaleuca* forest and selling it for firewood.

Transmigrants' relationships with local villagers revolved around land and cultural differences. In most cases, local people lost land to transmigrant settlements for little compensation. Sometimes locals resented the newcomers and would fight hard to have their land restored; at other times the settlements, with their improved transport access, were welcomed. Migrants' own behavior was often crucial. During the violence in West Kalimantan in 1998 between Madurese migrants and Dayaks, the locals were careful to distinguish between Javanese transmigrants, with whom they had no quarrel, and Madurese, perceived as a threat to Dayak identity and territorial control (Peluso and Harwell 2001: 107–8).

After 1989, tree crop schemes were gradually substituted for the largely unsuccessful food crop-based projects, and transmigrants became cash crop smallholders (first for rubber, then increasingly for oil-palm). While the initial properties were government-run, private company plantations were encouraged and were guaranteed a transmigrant workforce. Pulp plantations were also guaranteed Javanese labor under the HTI-Trans[5] scheme. Each household had to maintain 20 ha of tree plantation, which translated into considerably more cleared land than in an agricultural settlement (Fearnside 1997). The HTI-Trans scheme was abandoned in 2000, many of the settlers having a difficult time surviving in remote forest areas. With the movement of transmigrant smallholders into plantations the associated deforestation was attributed to those enterprises rather than to the transmigrants themselves.

Migration to the Central Highlands in Vietnam

In September 1970, the Ministry for Development of Ethnic Minorities in South Vietnam reported a population of 1.3 million people in the Central Highlands, 850,000 from the indigenous minority groups and 450,000 lowland Vietnamese.[6] Recent figures suggest just one million people from the local minority groups in a Central Highlands population of 4 million

(*Human Rights Watch* 2002). How has this demographic shift occurred and what have been its social and environmental results?

After the 1976 reunification of Vietnam the government relocated large numbers of lowland Vietnamese into upland areas. One million people were moved from the Red River delta to the provinces of Son La, Bac Thai, and Lai Chau, in the northern mountains. The focus then shifted to the Central Highlands, especially the provinces of Lam Dong and Dak Lak. Much of the early migration was due to Central Government programs which established state forest enterprises and coffee and rubber plantations. Since Vietnam's economic policy reform or *Doi Moi* (1986), government-organized movement has decreased, but spontaneous flows have intensified. The new settlers included not only lowland Kinh but minorities such as Hmong and Dao from the Northern Mountains. During the peak period of spontaneous migration between 1991 and 1996, over 32,000 migrants arrived in Dak Lak per year. Movements from 1976 to March 1999 included 311,764 planned migrants and 350,828 spontaneous arrivals (Pham 1999). Planned migrants received minimal government assistance (free transportation, housing, and basic supplies), but nothing was offered to those who resettled unofficially. Half the spontaneous migrants eventually returned to their communities of origin soon after arrival (Dang *et al.* 1997: 319).

In 1996 de Koninck observed that northern Hmong and Tay-Nung families settling temporarily on the margins of the pioneer areas cleared forest for recently arrived Kinh farmers (de Koninck 1996: 251). Local minority farmers had reluctantly begun clearing and selling their land to Kinh settlers, retreating farther into the forest where they continued to practice swidden cultivation (Pham 1999). The Vietnamese government eventually acknowledged that migration to the Central Highlands was contributing to loss of forests and serious environmental degradation. In April and May 1997, the Ministry of Agriculture and Rural Development decreed that all natural forests would be closed to commercial logging and all spontaneous migration into forest areas would be banned.

Tan (2000) reports of the determination of spontaneous migrant farmers at the coffee frontier as they carved a plot from the forest, claimed the land, found their own seed sources and entered the marketing networks. They achieved these goals basically by observing and questioning those already in the industry, even though they lacked previous experience with coffee. They financed the enterprise by growing short-term crops and selling their labor (Tan 2000: 57–60). Since that study, carried out in 1999, coffee prices declined steeply. Vietnamese domestic coffee prices plunged from US$3 per kg in 1995 to US$0.27 per kg in January 2002. Hardest hit by the crisis were ethnic minority farmers who had been squeezed out of their fallow land by the new settlers and also turned to coffee as an alternative to swidden. In February 2001 protests broke out in the Central Highlands and thousands of people demanded their lands back (*Human Rights Watch* 2002). Attempts have been made to change to alternative crops such as cotton, cocoa, or maize, but minority farmers remain in a critical position.

Itinerant migration into the forests

Gutta-percha collection

Many itinerant migrants into and out of the forests share a commonality; they interact with the forest yet do not replace it with permanent agriculture, the majority having an interest in its retention. These groups include commercial and nomadic gatherers, but also illegal loggers, bandits, guerillas, and refugees. They have some interests in common with indigenous long-settled communities whose stake in the forest is traditional and cultural as well as economic. Where they meet, immigrants and residents often compete for forest products. Some harvest forest resources without causing serious degradation; others degrade or clear-fell the forest through more intensive collection of non-timber forest products (NTFP) or timber when prices boom.

One such NTFP is gutta percha, a wild form of latex. It is extracted from twenty-three tree species of *Palaquium* and *Payena*, in the forests of Sumatra, the Malay Peninsula, Borneo, and southern Mindanao (Potter 1997). The latex was used to coat undersea cables in England, Europe, and the USA, and because for a long time no substitute existed, its value increased as cable networks expanded after 1870. The most famous collectors were the Iban, who were encouraged by the Brooke regime in Sarawak. Depending on the distance traveled, collectors usually stayed for one or two months in the forest, felling the trees on site and cutting incisions in the bark from which the latex oozed. Collecting gutta percha became an important rite of passage for young men, a way of proving their courage and valor against forest enemies while acquiring status and valuable trophies such as ceramic jars (Kedit 1993). Having been invited to Perak in the early years of the British establishment to look for gutta percha, they set up seasonal collecting routes along the rivers in the territories which became the Federated Malay States.[7] When authorities became alarmed at the rapid extinction of the trees and placed bans on collection, the Iban moved into the east coast sultanates of Kelantan and Terengganu, beyond the reach of colonial authority. From Sarawak they also penetrated into the neighboring territory of British North

Borneo, were rumored to have visited Mindanao, and were seen far up the Barito River in Dutch Borneo. French scientists eventually perfected a technique for extracting gutta percha from the leaves of the *Palaquium* trees, and the Dutch established plantations in Java to harvest leaves. Some Dayak groups, keen to secure their own supplies, also planted gutta trees near their longhouses.

Modern gaharu collection

Collection of gaharu incense wood is the contemporary equivalent to gutta percha collection, the former also stimulating itinerant migration into the forest. This fragrant heartwood species of *Aquilaria* is one of the few tradable items for indigenous Dayak living far from coastal markets in Borneo. Its collection from the forest, again usually undertaken by young men, also requires the felling of the trees, although not all *Aquilaria* trees contain gaharu (Eghenter, Chapter 11). Collection has traditionally been controlled through restriction of local access to particular forests and attempts to exclude outsiders. Dayak gaharu seekers generally check individual trees carefully, looking for signs of resin, and only cut those with favorable indications. Such systems break down when prices rise and powerful commercial interests invade the forests. Momberg *et al.* (2000) describe one such situation, in which city-based (and sometimes international) companies move gangs of professional collectors in and out by aircraft and helicopter. Such gangs are less conservative in their harvesting methods and are likely to cut all available trees (Eghenter, Chapter 11), forcing locals to change their behavior simply to compete.

Popularization of illegal logging

Following the fall of the Suharto regime, there has been an escalation of popularized illegal logging across Indonesia. The willingness and opportunity of the population to engage in logging stems from the breakdown of regulations imposed by the former government. Local people saw their traditional forests taken over by logging and pulp companies during the New Order period or were denied access to lands classified as protected forest. Many decided to take back those forests when opportunities arose (Potter and Badcock 2001). All across the forested areas of Kalimantan and in provinces such as Aceh and Riau in Sumatra, illegal logging has become a major occupation of young males, and at times entire families have relocated to live temporarily in the forests.

In recent fieldwork in Southern Riau, we discovered a well-constructed network at the village level (Potter and Badcock 2002). The economy of the village where we conducted research partly depended on rubber growing, but about 15 percent of households did not own rubber. Young males from one-third of the families were involved in logging, together with 200 young men from West Sumatra, who stayed temporarily in the village when not away in the forest. The main village elite who controlled the rubber trade also participated in timber marketing. A second-tier elite lent out chainsaws and financed the up-country expeditions. The operators and their assistants managed the month-long excursions into the forest for the felling and transportation of timber. Bribes to the police and forestry officials were factored into the equation. Operations were conducted in a nearby Nature Reserve, the boundaries of which were contested, and on the lands of a pulp and paper company. One young man recounted his experiences in a more distant forest from which he quickly retreated claiming large numbers of wild animals and "too many spirits" were providing too disturbing an environment. Villagers recognized that while this logging activity was unsustainable, it provided a good income opportunity while the supply lasted.

Interpreting the migration–deforestation link in Southeast Asia

Migrations into Southeast Asia's forests in the late nineteenth and early twentieth centuries reflected colonial commercialization, especially the development of export crops, and the political desire for increased state control over underpopulated "outlying" areas. The development of the Irrawaddy delta for export rice, the Deli region of East Sumatra for tobacco, rubber, and oil-palm, and parts of the Federated Malay States for plantation rubber, all led to migration. Such efforts involved extensive forest conversion. The examples presented above display contrasting migration processes, with different types of migrants involved who endured a variety of experiences on the forest frontier. Most of the migrations were government-assisted and did in fact result in the large-scale replacement of forests through cultivation, although many individual migrants failed for health or economic reasons. Itinerant movements were quite common.

Economic, political, and social factors were thus of greatest importance in these late nineteenth and early twentieth century examples. They therefore appear to be linked to the neo-classical deforestation explanatory framework, as discussed earlier. Above all, most of the schemes represented efforts by individuals to make money and improve their situation. At the same time, however, the state was the driving force in persuading or coercing people from remote places to join the plantation expansion initiatives. Thus the political ecology framework also contributes to explaining early plantation expansion and the link with migration. Particularly, it is the social ecology framework that can be used to explain why many of the migrants discussed above decided to leave the region once they had arrived there, or why they changed to an itinerant migratory behavior. The poor health conditions existing on the frontier in the nineteenth and early twentieth centuries resulted in many deaths. High mortality rates persisted despite colonial efforts to control malaria, a major scourge after forest clearance. The "contract coolies" would be replaced by another batch of laborers, but most survivors did not return home with any wealth. While spontaneous migration occurred in the locations of late nineteenth and early twentieth century plantation expansion, it did not have the dimensions of the spontaneous movements of the late twentieth century. Thus in these examples there is less need to rely on the neo-Malthusian or immiseration frameworks to explain migration into the forest zones.

Explanations of more recent migration schemes also need to rely on a combination of the four explanatory frameworks outlined above. The later movements were different from the early schemes, and primarily had a national development focus rather than simply involving individual profiteering. The justifications for these schemes included high population densities and the poverty of the migrating populations. Although this seems to suggest that neo-Malthusian-immiseration principles could be used to explain these early modern migration patterns, it is justifiable to question as to whether the government was the right agent to judge the poverty of the people and make decisions that they should migrate. As with so many other aspects of life during the Suharto era in Indonesia, government agencies used a fair degree of coercion to persuade poor people from Java, Madura, and Bali to join the transmigration schemes. Political security also played an important role in continuing the transmigration program. Army veterans, for instance, were settled along the borders with East Malaysia and Papua-New Guinea to better protect these boundaries. Both the schemes in Indonesia and Vietnam follow the same pattern of initial government-sponsored migrations to remote zones followed by significant spontaneous migration, suggesting that many people were desperate and tried to find better opportunities in the same locations.

Some care, however, needs to be taken when explaining spontaneous migration. Many spontaneous migrants are not so much looking for better opportunities; they may also be moving away from intolerable circumstances (see Bradley, Chapter 5; Limachi *et al.*, Chapter 3). Destabilization caused by war in neighboring territories increased the movement of refugee minorities into Thailand, especially in the north. While their impact on remaining forests has been considerable, they continue to be blamed for ecological problems stemming from previous land-use practices. They are now more involved in fruit and vegetable production than the shifting cultivation of opium (Delang 2002). Government policies encouraged movement into the northeast to counter forest-based communist insurgency between 1960 and the early 1980s; road- and railway-building and the granting of logging concessions facilitated settlement of the region, while dam construction pushed displaced settlers into the area (Missingham 1996). The movements of landless Thai lowlanders, responding to these opportunities and expanding dry-land farming and commercial cropping into the forests and uplands of both the north and northeast, were termed "spontaneous" (Uhlig 1984; Hirsch 1988). They were clearly influenced by government policy.

There is an equal need to rely on a social ecology interpretation for the Indonesian and Vietnamese cases discussed above. Economic failure was common in these schemes. Most of the food crop-based Indonesian transmigrant settlements were unsuccessful, resulting in individuals losing land, returning to Java, adopting indigenous production technologies or seeking alternative jobs, often in urban areas. Up to half the spontaneous settlers in Vietnam's Central Highlands either returned home or moved elsewhere soon after arrival.

The three cases of itinerant migration mentioned above, linked to the temporary exploitation of forest products, have had limited deforestation effects. However, the neo-classical, political and social ecology frameworks can be used to explain the processes involved. Migrants go in search of forest products due to market demand; in each of the three cases, the people who participated in the exploitation usually do not come from the poorest groups of society, nor do people from densely populated regions predominate among these itinerant migrants. The course of events in the case of gaharu extraction and the recent increase and popularization of illegal logging were profoundly influenced by the action of different tiers of government or other politically powerful actors. The discussion of gaharu exploitation presented by Eghenter in Chapter 11 of this volume clearly demonstrates the need to make use of a social ecology framework to understand the impact of migration with respect to this forest product.

Conclusions

This chapter and that of Andaya (Chapter 9) show the importance of forest-related migration since historical times. De Koninck and Dery (1997) have estimated that since the early 1950s, between 25 and 30 million people have migrated to pioneer areas in the Philippines, Indonesia, Vietnam, and Thailand. In the examples discussed above, migration and deforestation generally did result in the permanent alienation of the land to agriculture. Migration patterns have had many different causes, and these causes have changed over time. They coincided with new economic opportunities, but were also part of government policies. The need for people to move to the forest zones due to desperation seems to be a fairly recent trend.

Different types of migration have had and continue to have different levels of impact on the forest. Where migration leads to forest product extraction, the impact may be minimal. Even when people engage in agriculture at the forest frontier, their impact is a product of the land-use strategies they employ and their length of stay (Perz, Chapter 2).

Forest related migration and its deforestation effect can be explained using the three frameworks proposed by Wunder (2002): neo-classical, neo-Malthusian, and political ecology. To these three frameworks, I added a fourth: social ecology. The latter allows for the inclusion of factors such as unfamiliarity with the environment, the cultural and social traits of the in-migrants, and cultural differences between migrants and local groups, to explain how immigrants fared in their new area of settlement, and how that translated into forest conversion.

Notes

1. This word implies the movement of urban-based occupations into rural areas.
2. In 1885 Britain won the Third Burma War, giving it control over the whole country.
3. David Marr, in his preface to the 1985 translation of this work (p. viii), suggested that Tran Tu Binh's account of plantation life "may be assessed as exaggerated in tone, but reliable in substance".
4. Despite these trends, Indonesia initiated in 1995 the infamous "Million Hectare Rice Project" on deep peat lands in Central Kalimantan, which seemed to epitomise the worst features of earlier efforts. The project was cancelled after the fall of Suharto in May 1998, but not before the fires which engulfed the area in 1997 had emitted huge amounts of carbon into the atmosphere. The 15,600 families already settled failed in their efforts to grow rice and both they and neighbouring Dayaks saw their livelihoods destroyed.
5. HTI-Trans stands for "Hutan Tanaman Industri—Transmigrasi", ("Industrial timber plantation—transmigration").

6. This represented a rapid increase since 1953, when the lowland Vietnamese or *Kinh* people were estimated at a mere 30,000 (Hickey 1982: 303).
7. Perak, Selangor, Negeri Sembilan, and Pahang.

References

Adas, M. 1974. *The Burma Delta: Economic Development and Social Change on an Asian Rice Frontier, 1852–1941.* Madison: University of Wisconsin Press.

Adas, M. 1983. Colonization, commercial agriculture and the destruction of the deltaic rainforests of British Burma in the late nineteenth century. In R.P. Tucker and J.F. Richards eds., *Global Deforestation and the Nineteenth Century World Economy.* Durham: North Carolina University Press, 95–110.

Adhiati Adriana, S.M. and A. Bobsien. 2001. Indonesia's transmigration programme—An update. *Down to Earth*, July 2001.

Angelsen, A. and D.P. Resosudarmo. 1999. Krismon, *Farmers and Forests: The Effects of the Economic Crisis on Farmers' Livelihoods and Forest Use in the Outer Islands of Indonesia.* Bogor: Centre for International Forest Research.

Barber, C., E. Matthews, and D. Brown. 2002. *State of the Forest: Indonesia*, Washington: World Resources Institute, Global Forest Watch/Indonesian Forest Watch.

Brookfield, H.C. 1988. The new great age of clearance and beyond: What sustainable development is possible? In J. Denslow and C. Padoch eds., *People of the Tropical Rain Forest.* Berkeley and Los Angeles: University of California Press.

Colombijn, F. 1997. The ecological sustainability of frontier societies in eastern Sumatra. In P. Boomgaard, F. Colombijn, and D. Henley eds., *Paper Landscapes: Explorations in the Environmental History of Indonesia*, 309–340, Verhandelingen van het Koninklijk Instituut voor Taal-, Land- en Volkenkunde 178, Leiden: KITLV Press.

Dang, N.A., S. Goldstein, and J. McNally. 1997. Development and internal migration in Vietnam. *International Migration Review*, 31(2): 312–337.

de Koninck, R. 1996. The peasantry as the territorial spearhead of the state in Southeast Asia: The case of Vietnam. *Sojourn*, 11(2): 231–258.

de Koninck, R. and S. Dery. 1997. Agricultural expansion as a tool of population redistribution in Southeast Asia. *Journal of Southeast Asian Studies*, 28(1): 1–26.

Delang, C. 2002. Deforestation in Northern Thailand: The result of Hmong farming practices or Thai development strategies? *Society and Natural Resources*, 15: 483–501.

EIA (Environmental Investigation Agency) and Telapak Indonesia. 1999. *The Final Cut: Illegal Logging in Indonesia's Orangutan Parks.* London and Bogor.

ESCAP. 2001. *Asia-Pacific in Figures, 2001.* Bangkok, United Nations Economic and Social Commission for Asia and the Pacific.

FAO. 2000. *Forest Resources Assessment 2000.* Rome: Food and Agriculture Organisation.

Fearnside, P.M. 1997. Transmigration in Indonesia: Lessons from its environmental and social impacts. *Environmental Management*, 21 (4): 553–570.

Foxworthy, F.W. 1921. *Malayan Science Bulletin No 1—Commercial Woods of the Malay Peninsula.* Kuala Lumpur: FMS, Forest Department.

Fraser, A. 1998. Social, economic and political aspects of forest clearance and land-use planning in Indonesia. In B. Maloney ed., *Human Activities and the Tropical Rainforest Past, Present and Possible Future*, 133–150. Dordrecht: Kluwer Academic Press.

Geist, H. and E. Lambin. 2001. *What drives tropical deforestation? A meta-analysis of proximate and underlying causes of deforestation based on subnational case study evidence.* LUCC Report Series No 4, Louvain-la-Neuve: CIACO.

Government of Perak. 1911. *Annual Report.*

Hefner, R. 1990. *The Political Economy of Mountain Java.* Berkeley: University of California Press.

Hickey, G. 1982. *Free in the Forest: Ethnohistory of the Vietnamese Central Highlands, 1954–1976.* New Haven: Yale University Press.

Hidayati, D. 1994. Beyond government control? The adoption of indigenous practices by transmigrants in swamp and upland Kalimantan, Indonesia. Paper presented to *Asian Studies Association of Australia, Biennial Conference*, Murdoch University, Kalgoorlie, W.A.

Hirsch, P. 1988. Spontaneous land settlement and deforestation in Thailand. In J. Dargavel, K. Dixon, and N. Semple eds., *Changing Tropical Forests: Historical Perspectives on Today's Challenges in Asia, Australia and Oceania*, 359–376. Canberra: Centre for Resource and Environmental Studies, ANU.

Human Rights Watch. 2002. *Repression of Montagnards: Conflicts over Land and Religion in Vietnam's Central Highlands.* New York: Human Rights Watch.

Isager, L. and S. Ivarsson. 2002. Contesting landscapes in Thailand: Tree ordination as counter-territorialization. *Critical Asian Studies*, 34 (3): 395–417.

Kaimowitz, D. and A. Angelsen. 1998. *Economic Models of Tropical Deforestation: A Review*. Bogor: CIFOR

Kedit, P. 1993. *Iban Bejalai*. Kuala Lumpur: Ampang Press.

Kummer, D.M. 1992. *Deforestation in the Postwar Philippines*. Research Paper 234, Chicago: Department of Geography, University of Chicago.

Matthews, E. 2001. *Understanding the Forest Resources Assessment 2000*. Washington: World Resources Institute.

Missingham, B. 1996. Internal colonialism and environmental conflict: the case of Northeast Thailand. Paper presented to *Asian Studies Association of Australia, 20th Anniversary Conference*. La Trobe University, Melbourne, July 1996.

Momberg, F., R. Puri, and T. Jessup. 2000. Exploitation of gaharu, and forest conservation efforts in the Kayan Mentarang National Park, East Kalimantan, Indonesia. In C. Zerner, ed., *People, Plants and Justice: the Politics of Nature Conservation*, 259–284. New York: Columbia University Press.

Mubyarto. 2000. Indonesia's transmigration program: Experience and prospect. In K. Abe and M. Ishii, eds., *Population Movement in Southeast Asia: Changing Identities and Strategies for Survival*, 1–8. Osaka: Japan Center for Area Studies, National Museum for Ethnology.

Oekan, A. 1996. Social and environmental impacts of transmigration: A case study in Barambai, South Kalimantan. In C. Padoch and N.L. Peluso eds., *Borneo in Transition: People, Forests, Conservation, and Development*, 266–279. Kuala Lumpur, Oxford University Press.

Peluso, N.L. and E. Harwell. 2001. Territory, custom and the cultural politics of ethnic war in West Kalimantan, Indonesia. In N.L. Peluso and M. Watts eds., *Violent Environments*. Ithaca, Cornell University Press, 83–116.

Pelzer, K. 1948. *Pioneer Settlement in the Asiatic Tropics*. New York: American Geographical Society.

Pham T.M. 1999. Socio-economic analysis of shifting cultivation versus agroforestry system in the upper stream of Lower Mekong watershed in Dak Lak province. Unpublished thesis, MA in Economics of Development, National University HCMC and Institute of Social Study, The Hague, Netherlands.

Potter, L. 1997. A forest product out of control: Gutta percha in Indonesia and the wider Malay world. In P. Boomgaard, F. Colombijn, and D. Henley eds., *Paper Landscapes: Explorations in the Environmental History of Indonesia*, 281–308. Leiden: KITLV Press.

Potter, L. and S. Badcock. 2001. *The Effects of Indonesia's Decentralisation on Forests and Estate Crops in Riau Province: Case Studies of the Original Districts of Kampar and Indragiri Hulu.* Reports on Decentralisation and Forests in Indonesia, case studies 6 and 7, Bogor: CIFOR.

Potter, L. and S. Badcock. 2002. Impacts of rubber, oil palm and timber markets on farmer behaviour in forest margins of Riau, Sumatra. Paper prepared for International Symposium *Stability of Rainforest Margins in Indonesia*, Bogor, 29 September–4 October.

Rigg, J. and S. Nattapoolwat. 2001. Embracing the global in Thailand: Activism and pragmatism in an era of deagrarianization. *World Development*, 29(6): 945–960.

Rudel, T. and J. Roper. 1997. The paths to rain forest destruction: Crossnational patterns of tropical deforestation, 1975–90. *World Development*, 25(1): 53–65.

Stoler, A.L. 1985. *Capitalism and Confrontation in Sumatra's Plantation Belt, 1870–1979.* New Haven: Yale University Press.

Tan, S. 2000. Coffee frontiers in the Central Highlands of Vietnam: Networks of connectivity. *Asia Pacific Viewpoint*, 41(1): 51–67.

Tran T.B. 1985(1964). *The Red Earth: A Vietnamese Memoir of Life on a Colonial Rubber Plantation.* Translated by J. Spragens Jr, edited by D. Marr. Athens: Ohio University Center for International Studies.

Uhlig, H., ed. 1984. *Spontaneous and Planned Settlement in Southeast Asia.* Hamburg: Giessener Geographische Schriften Volume 58.

Wibowo, D. and R.N. Byron. 1999. Deforestation mechanisms: A survey. *International Journal of Social Economics*, 26(1,2,3): 455–474.

Wunder, S. 2002. Keynote address at the International Symposium *Land Use, Nature Conservation and the Stability of Rainforest Margins in Southeast Asia*, September 30–October 3, Bogor, Indonesia.

Wylie, E. 1993. Economic change and disease in Malaya c. 1820–1920: A study in human ecology. Unpublished PhD thesis, Griffith University.

9
Commodities, Culture and Migration in Early Modern Sumatra, the Malay Peninsula and Borneo

Leonard Y. Andaya

Introduction

The contemporary distribution of people in Southeast Asia and the denuded landscapes of once forested lands can be better understood by examining the history of resource use in the region. Relevant information on the root causes of modern phenomena, however, should be sought earlier than the statistical and descriptive data stemming from the records of the nineteenth century European colonial administrations and those of the succeeding independent Southeast Asian states. In the "early modern period" (c. 1500 – c. 1800) in Southeast Asia, migration played an important role in the progressive occupation and exploitation of the tropical forests and their resources. Understanding these early migration trends poses its own particular challenge, as prior to the nineteenth century records are spotty and reliant upon the chance observations of principally European or Chinese travelers, or on the occasional reports of officials of European trading companies.

Of the latter, it was the Dutch East India Company (VOC) that had the most extensive and systematic collection of documentation of seventeenth and eighteenth century Asia, the Middle East, and South Africa. From these reports, researchers have been able to extract a variety of information on resource use and migration patterns. However, much of this information is incomplete and even tentative due to the nature of the sources.

VOC officials compiled their reports, making use of information gathered by investigative missions sent by the Company, or through interviews with European and indigenous traders. Some of the topics in the reports were discussed without much depth or detail, thus researchers who make use of this documentation are often left to speculate as to the outcome of any investigation. It is important to emphasize this point to remind the reader of the highly tentative and speculative nature of some of the observations in the VOC reports. Nevertheless, through a careful contextual reading of the sources, informed by a careful application of ethnographic knowledge, scholars have succeeded in piecing together a useful picture of Southeast Asian communities in the early modern period. Occasionally local traditions have been compiled by indigenous scholars offering yet another glimpse of the past. Benedict Sandin (1967) for instance, an Iban himself, collected and studied the *tusut* or genealogies of the Iban people of Borneo.

Studies based principally on VOC and indigenous sources form the basis for the following observations regarding people's movements into and out of the forest in parts of Sumatra, the Malay Peninsula, and the island of Borneo in the early modern period. In particular I describe how the production of pepper in Sumatra and Southeast Asia and tin mining in Sumatra and peninsular Malaysia caused both forced and spontaneous migration into and out of the areas of production. The third example discussed is that of culturally-induced Iban migration into the forests of Borneo.

In section two I describe some of the social, cultural, and economic characteristics of the region that determined how migration became linked to resource exploitation. In section three I focus on early pepper production and migration patterns in Sumatra and South Borneo. Section four shifts to tin mining, looking at how prospecting for this metal has influenced movements of people. In section five I discuss the culturally motivated movements of Iban swidden agriculturists, and in section six I draw the themes together in a conclusion.

Forests and people in Southeast Asia before the early modern period

In many societies the forest is seen as dark, mysterious, and dangerous. The abode of strange and threatening creatures, it is regarded as the antithesis of "civilization". A shadow play (*wayang*) from Palembang, Indonesia, mentions a creature of the forest with no knowledge of Javanese language (*kawi*) or Javanese ways and hence a monster (*buta*) (B. Andaya 1993: 78). Among the Khmer people of Cambodia, the forest is viewed as a chaotic place from which civilization was born (Mabbett and Chandler 1995: 108). The frequently evoked imagery of "the primeval forest" conjures up the primordial, an association of the forest with the immense powers of creation. The forest, therefore, has also come to represent a source of spiritual strengthening and purification. Among early Christians in Europe, many devout individuals retreated to meditate in the forests to gain spiritual enlightenment, despite the fear and hostility that forests engendered in the minds of ordinary people (Harrison 1992: 62). In Theravada Buddhist societies there is a term "forest monks" indicating those monks who seek to purify and spiritually strengthen themselves by entering the forest and abandoning all contact with the material world ("civilization") (Mendelson 1975: 46–50; Swearer 1995: 130). The Malays view inhabitants of the forest as "powerfully different" because of their access to the supernatural forces associated with the interior jungles and hills (B. Andaya 1997: 3). For many, therefore, the forest has or had a sinister face but also one which promised self-strengthening through accessing its powers. These two aspects have both repelled and attracted individuals to the forest.

A second important point to consider in examining the relationship between the forest and people in Southeast Asia is the dearth of people in the region during the whole of the early modern period. Although Java is often cited as a heavily-populated island in Indonesia, most scholars attribute the population increase to Dutch colonial policies in the nineteenth century. In a Javanese court document from 1365, the *Desawarnana (Nagarakrtagama)*, there is reference to a heavily-forested landscape where wild animals roamed unchallenged (Robson 1995). Akin Rabhibidhana, a Thai anthropologist, was the first to alert scholars to the fact that much of the attention of the Siamese kings was devoted to maintaining and controlling human resources (1969). When historians and others began to examine the records more carefully, it became clear that for many Southeast Asian rulers it was not land but people that mattered.

According to the *Sejarah Melayu* (The Malay Annals), whose earliest known version dates from 1612, one of the first commands issued by the ruler after being ousted from the city of Malacca by the Portuguese in 1511 was to gather his people together at a new site (Brown 1952). It mattered little that his capital city was occupied; there was land in abundance, and it was people that determined the wealth of the kingdom (Brown 1952: 191). People were required to cater to the basic needs of a ruler, produce or obtain local items for international trade that contributed substantially to a ruler's income, and for status purposes (L. Andaya 1981: 194). This is the reason that in the seventeenth century a Siamese king extended an invitation to a refugee Makassar prince and his followers to settle in his land (Gervaise 1971: 50). In the seventeenth and eighteenth centuries the Makassar and Bugis people from South Sulawesi in Indonesia were among the largest of the many groups moving throughout the Malay-Indonesian archipelago, and they were eagerly courted by a number of rulers, including those of Siam and the Malay kingdoms. A seventeenth century Makassar king, whose policies had led to the flight of his people, lamented the fact that he was a ruler with no subjects to look after his food, water, clothes, and horses. Even in the late nineteenth century the king of Kutai in southeast Borneo is said to have told a European traveler: "Me want Kutai make big country; want plenty people" (Knapen 2001: 11).

There were various factors contributing to the extraordinary numbers of people leaving their homelands in the seventeenth and eighteenth centuries. One was economic, as groups moved temporarily or permanently from the interior towards the coasts to gain greater access to the maritime international trade. In other cases the movement was generated by political upheaval caused by VOC attempts to monopolize certain products, particularly spices (L. Andaya 2002, 2000; Reid 1993). This monopoly policy led to large and small wars, the most important of which was the war fought against the kingdom of Gowa which embroiled the whole of South Sulawesi from 1666 to 1669. The war and the subsequent harsh policies imposed by a VOC-supported native overlord contributed to a steady emigration of people from the island in the post-war period and for almost a century thereafter (L. Andaya 1981). VOC wars elsewhere in the archipelago simply added to the phenomenon of refugees moving to seek sanctuary with a new lord. Furthermore, there were forced movements of people as the VOC sought slaves for its cities and agricultural export enterprises. The traffic in slaves became widespread as indigenous rulers themselves participated in this trade to gain workers to raise cash crops for the international market (Reid 1983; Niemeyer 2000). The involvement of native communities in the lucrative tin and pepper trade became a major stimulus for the organization of human resources, the clearing of the forest, and the establishment of new settlements in the interior.

Pepper, migration, and the forest in Sumatra and southeast Borneo

Sumatra was the largest source of the world's pepper from the seventeenth to the nineteenth centuries (Bulbeck *et al.* 1998: 10). Pepper plants thrived in the conditions on Sumatra, with its relatively poor soils, warm temperatures, and more than 2,500 mm annual rainfall. Cubeb pepper and wild uncultivated pepper were initially the only types of pepper exported from Sumatra. The preferred black pepper (*Piper nigrum*) was not mentioned by Ibn Battuta in 1355 (Gibb 1929), and it is only in the fifteenth century that Ma Huan (1996 [1433]) noted that it was being grown in Pasai in northeast Sumatra. The plant may have been introduced in the second half of the fourteenth century from southwest India, its place of origin. While pepper was always an important export commodity, the enormous increase in international demand in the early seventeenth century encouraged the expansion of the pepper gardens. The difficulty was that shifts in demand could not be immediately accommodated; pepper is a slow-maturing plant taking four years before it yields fruits, and six to seven before the first substantial harvest. Although a plant can continue producing for thirty years, in Sumatran conditions there is a gradual decline in the quantity of berries after the twelfth year due to the lack of nutrients in the poor soils (B. Andaya 1993: 43, 79; Bulbeck *et al.* 1998: 62). As a result, growers had to prepare more land for the new plants even before the old vines stopped producing.

Pepper-growing, therefore, required a great investment of labor not only for the clearing of forests but also for the nurturing of pepper plants. After the men had cleared the land and prepared the ground, the women and children were given the tasks of weeding and planting the prop-plants (*cenkering*) around which the pepper vines were wound. Once the plants began to produce, there were two harvests: the larger one between April and June, and the smaller from September to November. The pepper pods were then cleaned, dried, bagged, and stored for transport to the coast (B. Andaya 1993: 43, 78–9). Only the high profits gained from the trade would have convinced people to engage in a crop which could be destroyed overnight by pests, weather-related disasters, or rampaging elephants. To avoid such risks, pepper-planters may have followed a practice common among subsistence rice-growers in Southeast Asia, who physically separated their plots and planted at different times to stagger the period of the plant's maturity.

By the early seventeenth century, the great international demand for pepper encouraged the rulers in Sumatra to seek an expansion of land under pepper cultivation. This was not a simple matter as the number of subjects under a typical ruler was relatively small and the people were normally engaged in some form of agriculture or forest-gathering activity. The rulers found the solution in the use of slaves. Although slaves were omnipresent in many indigenous societies, their role was primarily to augment the retinue status of their owners, and only secondarily to promote the latter's economic interests. The VOC demand for slaves for their own purposes created an industry and a market for slaves, also supplying indigenous rulers with the much needed labor for the expansion of the pepper gardens.

The seventeenth century Palembang and Jambi rulers of southeast Sumatra bought mainly Bugis slaves to do the back-breaking task of clearing the forest. In addition to purchasing slaves, these rulers also sent their subjects out to raid, and in 1669 the Jambi ruler was presented with some 2,500 people seized in slaving expeditions. Palembang rulers sent their ships as far as Borneo to raid for slaves. The demand was for "strong young men", and the best slaves were considered to be those from the eastern Indonesian island of Butun and from the Mandar, Bugis, and Makassar populations of South Sulawesi. In the 1660s slaves were the principal cargo of Bugis and Makassar ships arriving in Palembang. Some eight vessels came annually during this time, with each ship bearing about forty slaves. The ruler of Palembang in the 1670s also sent missions to Butun and South Sulawesi to augment his supply of slaves, buying as many as 200 at a time. Many of these slaves were sent upriver to clear more land for pepper cultivation (B. Andaya 1993: 96–7).

The slaves joined the already substantial numbers of Minangkabau people who had emigrated from their homeland in the central Sumatran highlands (*darat*) to settle in the upper reaches of the various rivers flowing from the interior to the east coast. The movement out of the Minangkabau *darat* would have begun much earlier, and by the end of the sixteenth century there is clear evidence of the presence of Minangkabau in the upper Batang Hari River in Jambi. They settled mainly in the districts of Tanjung and Kuamang where they became involved in planting pepper. Each year they sent more than 100 rafts carrying pepper downriver to the coast to be sold on the international market, with each raft carrying some 150 picul (9,000 kg) of pepper. By the early seventeenth century, others went through the Tembesi area to Merangin and up into the Palembang highlands. By the middle of the century Merangin had become one of the principal pepper areas in Sumatra; the Dutch estimated that the population of the area was 200,000. Although this figure is far too inflated, the size of the population in this interior region was substantially larger than the capital of Jambi which, according to Dutch estimates, had only some 4–5,000 people (B. Andaya 1993: 89).[1] Another concentration of Minangkabau pepper planters was in upriver Indragiri. In 1641 this area sent some 300 bahar (54,000 kg) of pepper to Malacca. So well-developed was this community of Minangkabau pepper-planters that the whole upriver Indragiri region was given the name of Kuantan and regarded as a separate entity from the downstream kingdom. Similar movements occurred in the upriver regions of the Siak and the Kampar rivers (Barnard 1998: 57–8, 62).

The coastal areas of west Sumatra were the other favored destination of Minangkabau from the *darat* eager to participate in the pepper trade. The ideal site was found to be the level ground along the banks of a river, somewhat inland to avoid flooding, and with access to water transport. Evidence shows the presence of pepper gardens already in the early decades of the sixteenth century between Tiku and Indrapura. The growing areas were located a distance from the coast and required an extensive transport network with *perahu* or native boats bringing the pepper first to collecting points, from whence they were brought and stored in warehouses at a stapling port on the coast. After 1663 the VOC established a monopoly on the pepper trade along the west coast of Sumatra, offering low prices and demanding white rather than black pepper, a process which was far more labor-intensive. The disincentive to growing pepper was exacerbated in the 1680s by a glut of pepper in the European market. Large areas of the coastal plain formerly planted with pepper were abandoned and reverted to scrub (Dobbin 1983: 72–3, 79).

Northeast Sumatra was also planted with pepper. According to the *Suma Oriental,* written by the Portuguese Tomé Pires sometime between 1512 and 1515 (Cortesão 1990), Pasai exported some 8–10,000 bahar (1.4–1.8 million kg) of pepper annually. Much of this would have been grown in the interior of Pasai, where Pires noted that there were occurrences of quarrels over pepper between Pasai and the interior producers. There was also some pepper grown all along the northeast coast of the island, with Pedir producing 2–3,000 bahar (216,000 to 540,000 kg) a year, while a number of other settlements along the coast had smaller quantities. The other major area of production was in southern Sumatra in Lampung, where Pires mentioned some pepper grown in the interior of the Tulang Bawang River under non-Muslim rulers (Cortesão 1990: 140–4, 159).

In speaking of these non-Muslims in the interior of Lampung, Pires was referring to the Abung. There was only a small amount of pepper being produced in this area in the early sixteenth century. In the following century, however, there seems to have been a vast increase in production among the Abung. During the seventeenth century, the production of pepper in the Banten kingdom in West Java was 10,000 bahar (1.8 million kg), most of which was grown in Lampung. The large numbers of laborers required to produce such quantities of pepper were very likely from the Abung, whose homeland in the interior was characterized by the Dutch as *volkrijk* (populous). Based on a formula suggested by Reid (1999), the Lampung area may have had a population of more than 50,000 people.[2]

As with the Minangkabau, the Abung would have emigrated from the interior down the rivers to participate in the lucrative pepper trade. From about 1740 they emerged as aggressive settlers occupying the lightly populated forested lands in Lampung with the aim of planting pepper. This move may have been encouraged by the Dutch themselves, who established a post along the Tulang Bawang River to oversee the pepper production among the Abung. In 1760 a VOC official claimed that 52,000 young pepper vines had been planted along the Wai (River) Abung, and that rice-growing was also being expanded (B. Andaya 1997: 4, 10). This remark may refer to the common practice found also among the Minangkabau of clearing the forest for plots of land which would be used for the pepper vines intercropped with rice for the first two years (Barnard 1998: 56). In swidden agriculture two years would have been the optimal time for rice production before the nutrients produced by the burning would have been exhausted. Pepper vines, however, would continue to grow for at least twelve years with diminishing harvests.

The expansion of pepper-growing among the Abung involved continual forest-clearing. Based on the experience of the Iban in Borneo, the Abung may have selected not primary but secondary forest, or areas which had already been cleared once of large trees by previous settlers. But even this practice would in time require clearing virgin forest for further pepper planting. The VOC encouraged increased production of pepper, and although there were sufficient forested lands available, the preference for secondary forests brought some tension among competing groups seeking new pepper plots. By the 1780s the Abung were reported to be burning rival pepper gardens in Lampung and as far away as the Komering River. Abung migration continued into the nineteenth century moving down the Tulang Bawang River and its tributaries. The demographic changes wrought by this steady flow of Abung were evident in the fact that a century later the term "Abung" was applied to any inhabitant in north, central, and eastern Lampung (B. Andaya 1997: 10, 12). Much of the virgin forest would have been cleared in the process of establishing these pepper gardens.

Pepper was also an important cash crop in southeast Borneo in the seventeenth and eighteenth centuries. In order to participate in the pepper trade, many Dayak left the upriver areas to settle closer to the coast in the vicinity of Martapura and the lower Riam Kiwa and Riam Kanan Rivers. Some of those who moved were attempting to prevent the further incursion of those from Kutai and Pasir in east coast Borneo into the pepper-growing regions; others sought greater economic opportunities in downriver Banjarmasin where there was strong demand for traders, coolies, and craftsmen. The ruler of Banjarmasin contributed further to the flow of Dayak southward by encouraging many of his upriver Dayak subjects to move to Martapura and Banjarmasin (Knapen 2001: 79). To participate in supplying the growing world demand for pepper, the sultan had to encourage his people to open new pepper gardens, as well as to engage in other tasks essential for such trade.

The generally favorable response by the Dayak to the ruler's appeal reflects the perceived economic advantage of such a move far more than any direct pressure by the sultan.[3] There is evidence of the shift of economic activity among some Dayak from hunting and gathering to rice-growing, presumably to supply the growing needs of a burgeoning population on the coasts. The Dutch, too, tried to persuade the Ot Danum Dayak to abandon the upriver areas to settle in downriver villages, even preventing the trade of salt up the Kahayan and Kapuas Rivers to hasten the process. Both the sultan's and the Dutch efforts were not wholly successful because of the Dayak's previous experience with coastal inhabitants. Disease in the coastal areas erupted periodically, and the economic and political ambitions of both the Malays and the Europeans on the coast contributed to uncertainty and even conflict. The interior by contrast seemed a haven (Knapen 2001: 95, 100–1, 390).

Despite the fear and uncertainty of the downriver world, there were many Dayak who did move to become pepper planters, rice-growers, and laborers. In 1800 from a population of around 100,000 in southeast Borneo, some 60,000 were estimated to be engaged in pepper-growing. The limited space available in the Banjar lowlands forced the planters to seek new plots toward the coast in the area of Tanah Laut and inward toward the Meratus Mountains. In the pepper-planting cycle, lands were eventually abandoned to become grasslands and pastures for buffaloes and deer. The decline in pepper production by the late eighteenth century encouraged some of the Banjar people who had been previously engaged in pepper cultivation to move to the lake (*danau*) region to grow wet-rice (*sawah*). This move coincided with a large migration of interior Dayak into the same region with precisely the same intention. Fortunately, the area was sufficiently extensive to absorb both groups, and in time the Dayak became Banjar. So rapid was the penetration of the lake region that by the 1830s all the available land had been occupied with a high population density (Knapen 2001: 390, 397).

The allure of profits from international trade affected even those groups that remained in the interior and were not engaged in pepper-planting. The first recorded contact between the upriver hunter-gatherer Punan Murung with the downriver Ot Danum Dayak was in the second half of the eighteenth century. In this period trading posts were established near limestone caves, which provided the Punan with temporary shelter and access to birds' nests for the China trade. As the Punan became increasingly involved in the collection of forest products, they decreased their hunting and gathering activities to rely for their subsistence on cultivated tubers, fruit trees, and sago palms. The income from the trade of forest products was then used to purchase cloth, tobacco, and salt. The tendency to adopt a sedentary lifestyle continued under the Dutch colonial government, and the last groups following a nomadic existence had ceased to do so by the first half of the twentieth century (Knapen 2001: 90; Dounias and Leclerc, Chapter 7).

Tin, migration, and the forest in Sumatra and the Malay Peninsula

Besides pepper, tin was the other major commodity in the early modern period that had a significant impact on both migration and the forest environment. Tin recovered in Southeast Asia before the nineteenth century was mostly alluvial—fragments from rock deposits which had settled in riverbeds. Because of the nature of the deposits, the tin could be reached in pits only a few meters deep. One of the earliest methods used to recover tin was through panning, or *dulang* washing. After heavy rains, alluvial tin was washed down into the streams where the Malays would then use these pans to separate the ore from other objects. Another way of obtaining tin was through the *lampan* method, in which a ditch was dug near a stream into which earth was thrown. The earth was then stirred to break up the clay and separate out the tin sand. The method which had a greatest impact on the environment prior to the nineteenth century was mining. In the river valleys where many of the tin mines were located, the surface was cleared of rocks and large trees, including the roots. Shafts were then dug to about 2 to 2.5 m, and the tin found was washed away from the sand using water conveyed by hollow logs to the mines. If no springs were available in the vicinity, wells were dug to provide the water. For the Malays it was a secondary occupation which occurred after the rice harvest. Men, women, and children worked the mines from 5 a.m. to about 10 a.m., and again from 3 p.m. to 6 p.m. Tin-collecting activities were irregular since they were disrupted for rice-growing and for required corvée labor for the ruler (B. Andaya 1979: 395–7; Heidhues 1992: 8).

Early tin mining was located in the western half of the Malay Peninsula and the southeastern portion of Sumatra. Prior to the nineteenth century, the tin-mining areas were far less extensive than the present-day, but Perak on the Malay Peninsula was still the major producer. The Kinta Valley, home of the largest tin fields in the world, was largely untapped at the time, and Larut, another important later site of tin mining, produced only a small amount before the nineteenth century. The mines were centered at the base of foothills and by the sides of major rivers. A VOC report from 1768 lists the tin mining areas as being mostly along the Perak River and its tributaries, and in the districts around Kuala Kangsar and present-day Kelian Indah (B. Andaya 1979: 395). The major tin mines in Sumatra were limited to a 15 km stretch between the Tapung Kiri and Kampar Rivers, but by 1687 new mines were being opened up in upriver Rokan. The entire region could produce 4–500 bahar (68,000–85,000 kg) of tin annually (Barnard 1998: 64, 80).

Tin was a useful commodity for the VOC because it could be exchanged for spices in India to purchase Indian cloth, the most highly-desired trade item in the Malay-Indonesian archipelago. Tin first appeared in VOC auctions in 1667 and 1668, and the Dutch post in Malacca was instructed to collect between 100,000 to 400,000 lbs (45,500 to 181,800 kg) annually for the European market (L. Andaya 1975: 106; Glamann 1958: 19; Lewis 1970: 78–9). China, however, was the destination for as much as 75 percent of the tin mined in the Straits region, thus giving some indication of the large demand for tin in the seventeenth century (Barnard 1998: 64). This demand brought an influx of people seeking to participate in the tin trade. New migrant villages were founded in Sumatra, mainly by Minangkabau settlers descending to the upper reaches of the major rivers from their relatively heavily-populated homeland in the central highlands. Interior depots arose at Patapahan on the Tapung Kiri River, as well as in Kabon, Kota Rena, Bangkinang, Air Tiris, and Salo on the Kampar River, to provide services to the mining villages and to facilitate the sale of tin. Traders came upriver with salt, cloth, specie, and slaves to exchange for tin. If the tin-producers were unable to obtain acceptable prices, they used the footpaths linking the Siak and Kampar Rivers to reach the Rokan River where traders from Aceh were willing to offer better prices (Barnard 1998: 80, 88–9). There was the further option of going to Johor-controlled downriver Kampar to exchange their tin for salt and cloth. The VOC, too, had their own agents competing with the others for a part of the tin trade (L. Andaya 1975: 147, 204). But Sumatra was only a minor tin producer. The bulk of seventeenth and eighteenth century tin came from the Malay Peninsula. Along with the areas in Perak, tin was found further south in Selangor, Linggi, Kelang, and Sungai Ujong.

One of the consequences of the growing importance of the tin trade was the arrival of two outside groups into Sumatra and the Malay Peninsula in the seventeenth and eighteenth centuries: the Bugis and the Chinese. The Bugis constituted the largest of four major ethnic groups in South Sulawesi. After the Makassar War of 1666–9 fought between the Makassar kingdom of Gowa and the VOC assisted by its ally the Bugis, there was a steady stream of refugees leaving the island. Initially the flood was a result of fear of retribution, but in later years the flow continued as individuals and groups led by their leaders sought to escape the arbitrary and oppressive deeds of the Bugis overlords of Bone backed by VOC power. These refugees met varying fates in the archipelago in their attempts to find a new lord to serve. The most successful eventually settled in the thinly-populated lands of the ruler of Johor on the west coast of the Malay Peninsula.[4] It was no coincidence that the Bugis selected lands that were noted for their tin production. Once settled with their sizeable following on the west coast of the Peninsula, they successfully engaged in the lucrative tin trade. In 1714 the Dutch named the Bugis-controlled settlements of Selangor, Linggi, and Kelang on the Peninsula, as well as Siak on Sumatra, as the most important tin producers of the kingdom of Johor (L. Andaya 1975: 239). The sources do not reveal whether it was the Bugis themselves who panned or mined the tin, but they were the traders of this ore to the outside world.

There are no population figures for the Bugis communities in the Peninsula, but the same 1714 Dutch report lists as part of the Johor fighting force some 200 soldiers from Linggi and Sungai Ujong, inhabited mainly by Bugis and Minangkabau, and 300 soldiers from Selangor, inhabited mainly by Bugis. Another 300 soldiers could be mustered from Kelang, which is listed mainly as a Minangkabau place, though it was one of the major Bugis-controlled tin areas (L. Andaya 1975: 333). Even these figures may, however, have grossly underestimated the numbers of Bugis in the region. When the Bugis leader Daeng Marewa from Linggi assisted the Johor ruler in regaining his kingdom from Minangkabau invaders in 1721, he confidently told the Dutch that Bugis territory stretched from Malacca to Bernam. Although some eighty soldiers from the ruler of Bone in Sulawesi were among the Bugis forces used against the Minangkabau, the vast majority were those recruited from the Bugis communities on the Malay Peninsula (L. Andaya 1975: 289–90). The presence of such large numbers of Bugis would have meant greater clearing of forests for homes and for agriculture. Depending on the suitability of the land, they would have attempted to introduce some agricultural activity, notably rice-growing, in their new homeland.

The other major group that came to Sumatra and the Malay Peninsula in the eighteenth century to participate in the tin trade was the Chinese. Although Chinese traders were prominent in the area for centuries, the first major influx of Chinese occurred in response to the rising international demand for tin. With the introduction of Chinese tin-miners to the island of Bangka off the coast of southeast Sumatra in the early eighteenth century and to Perak from 1777, Chinese mining techniques were brought for the first time to Southeast Asia. The method most favored by the Chinese in Bangka was termed the *kolong* (pit). To find a suitable site the Chinese used a metal hollow bore about 6 m long to bring samples of the soil to the surface. From these samples they were able to determine the quantity and quality of the tin deposit. Once a site was selected, a large number of men in teams of twenty-five to thirty were actively employed in the mine. A nearby stream was dammed to divert water to a conduit which ran a waterwheel to pump water out of the mine. The workers were thus able to work as deep as 6 m, using tree-trunk ladders to climb out of the pit with baskets of soil balanced on shoulder-poles. The water channeled to run the wheel was also used to remove the topsoil and wash the ore. The largest mines could measure as much as 30 m in length and width. Efficient workings of such a mine required a fairly large number of men, from the miners themselves to carpenters and smiths. In addition, the Chinese method of smelting was labor-intensive. As their local counterparts, the Chinese tin-miners used charcoal for the smelters but employed a larger furnace with stronger bellows made from a tree trunk and worked by men on each end who rotated in teams of six. The smelting occurred once a year at a central location where the tin was formed into ingots of a standard shape and weight. The ingots were then transported in wooden barrows (Heidhues 1992: 11–15). This method of mining tin which originated in China would have also been used by Chinese miners on the Malay Peninsula.

Chinese methods of tin mining and smelting required a far greater investment of labor and resources than the methods used by the Malay communities. For the local population, the principal method for the recovery of tin was panning in the streams. Moreover, it suited their lifestyle in which tin was regarded as a supplementary activity to be engaged in by the whole family after the principal rice harvest. While there were some who actually dug mines and worked more intensively to extract and smelt the tin, these methods were on a smaller scale than the Chinese operations. Chinese tin mining technology required a large investment of capital, provided mainly by the local rulers and the VOC. The funds were necessary to recruit Chinese workers, purchase the necessary equipment, and obtain sufficient supplies until the first ingots could be sold (Heiduhues 1992: 15). Unlike the Malay tin-miners, the Chinese were employed full-time in the process of identifying suitable sites for mines, extracting, smelting, and selling the tin. The landscape was thus pocked with abandoned mines, tailings, and denuded forestlands which had provided the timber supports for the mines, the troughs to convey the water from the streams to the mines, and the fuel to feed the furnaces. The large number of Chinese workers engaged full-time in the mines also encouraged the growth of towns to service these miners. These Chinese settlements grew into the principal towns and cities on the west coast of the Malay Peninsula.

Iban migration and the forest in Borneo

Another historical example of the relationship between people movements and the forest is the case of Iban migrations in Borneo. It was not pepper, tin, or any identifiable economic incentive which contributed to this phenomenon, but a more cultural motivation. According to their *tusut*, or sung genealogies, the Iban began moving sometime in the sixteenth century from their ancestral lands in the Batang (River) Kapuas region into the Sarawak river systems of the Saribas, Lupar, and Batang Air. The principal reason for the moves appears to have been the desire to become pioneers in the opening up of new lands, and indeed a major aim of the *tusut* was to legitimize claims of land ownership (Sandin 1967: 3). In reconstructing the vast prehistoric migrations of the Austronesian speakers (the ancestors of most of the inhabitants of archipelago Southeast Asia, including the Iban) from Taiwan southward to the islands of the Malay-Indonesian archipelago, Bellwood (1996) suggests that the pattern was to settle the easily accessible coastal areas first before moving into the interior. Moreover, the rapid expansion of the Austronesian speakers was encouraged by "founder rank enhancement," referring to the quickest way to rise in status through becoming a founder of a new community (Bellwood 1996). Pringle attributed precisely this motivation to the Iban migration, which spread relatively quickly from the Kapuas sometime in the mid-sixteenth century into the river systems of the Second Division and then into the Third Division of the Batang Rejang in Sarawak by the time of the arrival of James Brooke in 1841 (1970: 12, 41).

According to the *tusut* compiled by the Iban ethnographer, Benedict Sandin (1967), each move was associated with the opening up of a new area by a warrior pioneer. The clearing of the land was regarded as part of the task of the warrior, battling to domesticate the unrestrained forces of the forest to make the place safe for human habitation. The Iban did not penetrate the deepest forest, instead moving generally from the mouth to the middle reaches of a river not too distant from the sea. The Iban usually cleared the forest along the river banks, where they then planted their rice. The land cleared by a warrior pioneer was divided among his followers, and the group occupied a few temporary houses before eventually settling in a permanent longhouse. With a relatively large population and a fierce reputation as warriors, the Iban easily displaced the hunter-gatherer communities by absorbing them or forcing them to move further upriver. Iban migration, however, was not unidirectional, and there is evidence that in about 1800 there was a reverse movement from the middle and upper Batang Lupar in Sarawak back into the Kapuas (Pringle 1970: 42).

Iban migration had a significant impact on the forests of Sarawak. With the enhanced status achieved by opening new lands, ambitious individuals had a strong cultural incentive to move into virgin forest or even into areas occupied by other ethnic groups. More than one site was cleared before the group eventually decided to build the permanent longhouse. After a time, others within the group would break away seeking to open their own communities, necessitating the clearance of more land. Although there was a preference for areas that had been cleared by earlier groups and had reverted to secondary jungle, the number of these areas was limited, forcing later groups to resort to clearing more virgin forest. As the Second Division in Sarawak was relatively densely populated, it follows that there would have been a faster rate of forest-clearance to accommodate the Iban practice of swidden agriculture (Pringle 1970: 35). Even though Sandin (1967) used only the *tusut* from Paku, his home river, the movement of this group alone was very extensive. Other groups would have undergone a similar process, contributing further to a fairly substantial exploitation of the forestlands in Sarawak between the sixteenth and nineteenth centuries.

Conclusions

In the history of Southeast Asia, the forest has always had a dual aspect. On the one hand it is regarded as an unknown space—dangerous, wild, uncontrolled, and hence "uncivilized". On the other hand, perhaps because of its "unknown" quality, it is viewed as unblemished, pure, and thus spiritually powerful. For Southeast Asians, therefore, the forest has been both feared and respected. There have always been movements into and out of the forest for political and religious reasons. Those seeking to escape the hand of government sought refuge in the forest, while heroes and holy men and women entered it to gain spiritual strengthening. In traditional literature, it is the period spent in the forest which results in spiritual purification and power, enabling the individual to return to "civilization" triumphant. Though this form of in- and out-migration continues to occur in Southeast Asia, it is by nature small and tends to leave the forest environment relatively undisturbed.

Cultural factors have also played a part in the movements of people in and out of the forest, such as the case of Iban migrations in Borneo. Rank enhancement through the founding of new settlements, coupled with the challenge of conquering the forest, has encouraged a rapid spread of the Iban peoples along the rivers of Borneo. Yet the Iban have never moved very far inland, tending to favor the lower reaches of the river where they can move in and out of the forest. These Iban movements are aimed at domesticating the wilderness through the establishment of settlements, thus causing some change to the forest environment

The greatest penetration and intrusion on the forest has been the result of economic development. In the early modern period, world demand for pepper and tin encouraged states and individuals to clear the forest and to use its resources to facilitate the planting of pepper and the extraction of tin. In the case of pepper, large forested areas were beginning to be cleared in parts of Sumatra, Malay Peninsula, and Borneo for the plantations. Equally intrusive were the movements of laborers, both free and enslaved, to work these pepper gardens. The nature of the pepper plant resulted in a steady clearing of new land to maintain productivity. Tin did not require as much cleared space, but the introduction of Chinese methods of extraction and the establishment of Chinese mining settlements had a major impact on the forest. After the initial clearance of the forest, the land never truly reverted to secondary forest because many of the cleared areas were used for other new profitable commodities in the late nineteenth and twentieth centuries. Moreover, as clearing penetrated further and further into the interior, roads and railways were needed to allow transportation of the products to port. This was a process that had begun on a smaller scale in the early modern period, accelerating since the late nineteenth century in many parts of Southeast Asia.

As the forests of Southeast Asia become depleted, there is a corresponding dwindling of their mystery and hence their awe. They are now regarded as the last frontier to be economically exploited by the few, rather than a resource which needs to be preserved for the wellbeing of the whole community. What this means is that the state and powerful commercial interests will control movements into these few remaining forested areas in order to gain the maximum economic benefits before the resource totally disappears. Those who persist in maintaining a lifestyle dependent on the forests will be "encouraged" to leave their forest domain and become a progressive and modern member of the national society. This process is radically different from the approach of the early modern kingdoms in Southeast Asia. Then, the desire was to retain the forest peoples in their environment because their skills as gatherers of forest products were essential for the economic success of the downstream/coastal kingdoms. The introduction of foreign crops into the forested environment challenged but did not displace the economic value of the forest inhabitants. It would seem that the only way the remaining forests and those who inhabit them can be preserved is by finding a way of returning to the economic complementarity between those dwelling within and outside the forest borders.

Notes

1. The Minangkabau homeland has rich rice-growing lands which have supported substantial populations. When Stamford Raffles traveled to

the Minangkabau highlands in the early nineteenth century, he firmly believed that within a fifty mile radius of Pagar Ruyong was a population of not less than one million (Raffles 1991: 363).

2. Reid calculated that in the eighteenth century a family pepper plot could produce an average of 200 kg of pepper per year. With the weight of pepper at the time being one bahar to 180 kg, to produce the 10,000 bahar of pepper would require some 8.561 families. With the family size calculated at about six, there would therefore have been a population of somewhere around 51,370 people (Reid 1999: 125).
3. The closer the subjects were to the court, the easier were they counted and "seen" hence controlled (Scott 1998). The Dayak would have understood this, but the economic prospects of such a move obviously outweighed the disadvantages of being subjected to greater control from the center.
4. For a discussion of the internal situation in Sulawesi which contributed to the flight of the Bugis and other ethnic groups, see L. Andaya 1995.

References

Andaya, B.W. 1979. *Perak: The Abode of Grace; A Study of an Eighteenth Century Malay State.* Kuala Lumpur: Oxford University Press.

Andaya, B.W. 1993. *To Live as Brothers: Southeast Sumatra in the Seventeenth and Eighteenth Centuries.* Honolulu: University of Hawaii Press.

Andaya, B.W. 1997. Raiding cultures and interior–coastal migration in early modern Island Southeast Asia. In B. Barrington, ed., *Empires, Imperialism and Southeast Asia*, 1–16. Monash: Asia Institute.

Andaya, L.Y. 1975. *The Kingdom of Johor 1641-1728.* Kuala Lumpur: Oxford University Press.

Andaya, L.Y. 1981. *The Heritage of Arung Palakka: A History of South Sulawesi (Celebes) in the Seventeenth Century.* The Hague: Martinus Nijhoff.

Andaya, L.Y. 1995. The Bugis-Makassar diasporas. *Journal of the Malaysian Branch of the Royal Asiatic Society*, 68(1): 119–138.

Andaya, L.Y. 2000. Unravelling Minangkabau ethnicity. *Itinerario, European Journal of Overseas History*, 24(2): 20–43.

Andaya, L.Y. 2002. The trans-Sumatra trade and the ethnicization of the "Batak". *Bijdragen tot het Taal-, Land- en Volkenkunde*, 158(3): 367–409.

Barnard, T.P. 1998. Multiple centers of authority: Society and environment in Siak and Eastern Sumatra, 1674–1827. PhD Thesis, University of Hawaii, Honolulu.

Bellwood, P. 1996. Hierearchy, founder edeology and Austronesian expansion. In J.J. Fox and C. Sather, eds., *Origins, Ancestry and Alliance: Explorations in Austronesian Ethnography*, 18–40. Canberra: The Australian National University, Department of Anthropology.

Brown, C.C. (tr.). 1952. *Sejarah Melayu or "Malay Annals"*. Singapore: Malaya Publishing House.

Bulbeck, D., A. Reid, L.C. Tan, and Y. Wu, eds. 1998. *Southeast Asian Exports since the 14th Century: Cloves, Pepper, Coffee, and Sugar.* Leiden: KITLV Press.

Cortesão, A. (tr. and ed.). 1990. *The Suma Oriental of Tomé Pires*. Vol. 1. New Delhi: Asian Educational Services.

Dobbin, C. 1983. *Islamic Revivalism in a Changing Peasant Economy: Central Sumatra, 1784–1847.* London: Curzon Press.

Gervaise, N. 1971. *An Historical Description of the Kingdom of Macasar in the East Indies*. Westmead: Gregg International Publishers Ltd.

Glamann, K. 1958. *Dutch Asiatic Trade, 1620–1740*. Copenhagen: Danish Science Press.

Harrison, R.P. 1992. *Forests: The Shadow of Civilization*. Chicago: University of Chicago Press.

Gibb, H.A.R. (tr. and ed.) 1929. *Ibn Battuta, Travels in Asia and Africa 1325–1354*. London: Broadway House.

Heidhues, M.F.S. 1992. *Bangka Tin and Mentok Pepper*. Singapore: Institute of Southeast Asian Studies.

Knapen, H. 2001. *Forests of Fortune? The Environmental History of Southeast Borneo, 1600–1880*. Leiden: KITLV Press.

Lewis, D. 1970. The Dutch East India Company and the Straits of Malacca, 1700–1784. PhD thesis, Australian National University, Canberra.

Mabbett, I. and D. Chandler. 1995. *The Khmers*. Oxford: Blackwell.

Ma Huan. 1996 (1433). *The Overall Survey of the Ocean's Shores*. Bangkok, White Lotus Press.

Mendelson, E.M. 1975. *Sangha and State in Burma*. Ithaca: Cornell University Press.

Niemeyer, H.E. 2000. Slavery, ethnicity and the economic independence of women in seventeenth century Batavia. In B.W. Andaya, ed., *Other Pasts: Women, Gender and History in Early Modern Southeast Asia*, 174–94. Honolulu: Center for Southeast Asian Studies.

Pringle, R. 1970. *Rajahs and Rebels: The Ibans of Sarawak under Brooke Rule 1841–1941*. Ithaca: Cornell University Press.

Rabibhadana, A. 1969. *The Organization of Thai Society in the Early Bangkok Period, 1782–1873*. Data Paper No. 74. Ithaca: Southeast Asia Program.

Raffles, S. 1991. *Memoir of the Life and Public Services of Sir Thomas Stamford Raffles*. Singapore: Oxford University Press.

Reid, A. ed. 1983. *Slavery, Bondage & Dependency in Southeast Asia*. St. Lucia: University of Queensland Press.

Reid, A. 1993. *Southeast Asia in the Age of Commerce 1450–1680*. Vol. 2: "Expansion and Crisis". New Haven: Yale University Press.

Reid, A. 1999. Economic and Social Change, c. 1400–1800. In N. Tarling, ed., *The Cambridge History of Southeast* Asia, 116–63. Vol. 1, Part 2. Cambridge: Cambridge University Press.

Robson, S. (tr. and ed.). 1995. *Desavarnana (Nagarakrtagama)* by Mpu Prapañca. Leiden: KITLV Press.

Sandin, B. 1967. *The Sea Dayaks of Borneo before White Rajah Rule*. London: Macmillan.

Scott, J.C. 1998. *Seeing like a State*. New Haven: Yale University Press.

Swearer, D.K. 1995. *The Buddhist World of Southeast Asia*. Albany: Suny Press.

10
Migration and Tropical Forests in the Central Highlands of Vietnam

Tran Van Con

Introduction

Since 1976, when Vietnam was reunited, its government has attempted to extend the centrally-directed socialist system over the entire nation. From 1986 onwards there was a radical change from centrally-planned to market-driven economic growth programs. At present, the economic development policy focuses on four key strategies: 1) increasing human resources; 2) raising national scientific and technological capabilities; 3) developing infrastructure and key industries; and 4) forming market mechanisms under government management.

The government of Vietnam is particularly concerned with the development of the poor mountain regions and the ethnic minorities who live there. Efforts to promote fixed settlement and cultivation as a strategy to gradually improve the living conditions of shifting cultivators, and reduce forest fires and soil damage, were addressed in legislation such as Resolution 71-TW dated 23 February, 1963 and Resolution 38-CP dated 12 March, 1968. In 1980 the government promoted the movement of people from overcrowded areas in the lowlands to the under-populated highlands in the center and south of the country. Initially, however, the government only assisted in identifying new economic zones, providing land and basic housing, and supplying rice for three to six months. No technical advice or investment capital was made available. In later efforts, these policies were revised and better assistance was provided.[1]

Economic development policies and related migration have profoundly affected forest cover and forest conditions in Vietnam. Over the past few decades, the forested areas have both declined and increased and are now standing at 10.9 million ha. Forest conditions have generally declined.

Spontaneous migration in Vietnam, especially from the northern provinces to the south, has become an important social phenomenon in the two last decades, drawing the attention and concern of many government planners and policy makers. Before 1985, almost all of the migration that took place in Vietnam was the result of careful government planning (T.H. Nguyen 1998). After 1985, however, spontaneous movements have increasingly contributed to the migration flows. The most popular destinations for these spontaneous migrants have been the Central Highlands, in addition to the north-east of the Mekong Delta.

In this chapter I examine migration dynamics in the Central Highlands of Vietnam in order to demonstrate the social, ecological, and political profiles of immigrants into forested areas. I analyze the circumstances that led to the observed migration processes and their impact on the conservation of tropical forests in the region. I particularly focus on the factors that constitute the mechanisms that start with migration and end with forest destruction. I also discuss briefly the impacts of forest conversion, especially on the wellbeing of both migrants and original inhabitants in the Central Highlands. Finally, I outline a program of action that needs to be implemented if any progress is to be made in both reducing migration towards the Central Highlands and mitigating the negative impacts of recent migration processes.

The chapter is divided into five main sections. Section two, following this introduction, contains background information on Vietnam and its forests. In section three I describe the present situation in Vietnam, providing a profile of migrants and migration patterns. In section four I discuss in detail the impacts of migration into the Central Highlands. In Section five I propose a comprehensive set of measures that form the minimum requirements if any of the observed impacts are to be addressed. In the last section I draw some major conclusions.

Forests, economic development, and migration in Vietnam

Vietnam covers an area of about 33 million ha and has a relatively complicated terrain: countless mountains, rivers, highlands, and plains, and a stretching and meandering coastline that includes islets, bays, and large ports. The northern part of the country consists of mountains and hills that surround the central plains and deltas of the Red and Thai Binh Rivers. The central part is divided into many sections by rivers originating from western mountain ranges that have formed deep and narrow valleys. The southwestern side of this central region is a complex of granite highlands called "Tay Nguyen", with an average elevation of 900 m above sea level. The southern part of the country is dominated by the Mekong Delta plains.

Some 80 percent of the land area is mountainous and relatively sparsely populated. About 75 percent of the population of 80 million lives on intensively cultivated alluvial plains. Population densities vary between provinces, from twenty-six persons per km^2 in Lai Chau in the northern mountains to 1,200 in Thai Binh in the Red River Delta, with a national average of 280 persons per km^2. Almost 87 percent of the population is Kinh, who are Vietnamese-speakers. The remaining 10 million people belong to fifty-three minority groups. Most of the latter live in the mountainous interior of the country where they often constitute the majority at district or even provincial levels. The average annual population growth is about 1.7 percent. Currently, the country is administratively subdivided into sixty-one provinces and cities; the provinces are further subdivided into 622 districts which are again divided into 10,511 communes—the lowest level of the state administrative system. Each commune, finally, contains several villages.

The main forest types in Vietnam are: evergreen broadleaved tropical moist forest in the mountains; evergreen broadleaved tropical monsoon forest in the high mountains; semi-deciduous tropical monsoon forest; dry dipterocarp forests in the highlands; sub-tropical pine dominate forests; mixed conifer-broadleaved associations; and mangrove and bamboo-hardwood associations (N.P. Tran 1970; Thai 1972). About 12,000 species of plants are estimated to grow in Vietnam, of which only 7,000 have been described and up to 1,000 species are known to be endemic to the country. At least 1,000 tree species are large enough to produce timber.

Data from 1999 indicate a forested area of 10.9 million ha, or 33.2 percent of the country's area (Nguyen 2001). Some 9.4 million ha contain natural forests and 1.5 million ha are plantations (DFD 2001). From 1976 to 1995 about 2.8 million ha of natural forest in Vietnam was lost (Table 10.1). Areas under plantations partly compensated for the loss of natural forests. As a result, during the 1990s there has been a net increase in forest cover. The forest quality, however, has declined notwithstanding this forest cover

Table 10.1. Changing of forest area over time (1,000 ha)

	1943	1976	1980	1985	1990	1995	1999
Natural forest	14,300	11,076	10,186	9,308	8,431	8,253	9,444
Plantation	0	92.6	422.3	583.6	744.9	1,050	1,471
Total	14,300	11,169	10,608	9,892	9,176	9,303	10,915
Forest coverage (%)	43.0	33.5	30.8	30.0	27.8	28.2	33.2

Source: Forest Inventory and Planning Institute (FIPI 1996).

increase. There are many reasons for the reduction of Vietnam's forest, including wars, poor management, overexploitation of timber, fuel, and other resources, and forest clearance for the planting of rubber, coffee, food crops, and other agricultural purposes. Economic development policies and related migration have also profoundly affected forest cover and forest conditions in Vietnam.

The Central Highlands of Vietnam consist of four provinces: Kontum, Gialai, Daklak, and Lamdong. These provinces have a total area of 57,373 km^2 and a population of about 4.2 million. The region has a forest cover of 57 percent, the largest percentage forest cover in the country. The rates of forest loss and degradation, however, are very high. There are two main reasons for this degradation that directly relate to population pressure: the need for agricultural land to be used for food production, and the need for wood for a variety of purposes. Migration has played a major role in population increase in this part of Vietnam, and the related impact on forests.

The present situation and features of migration dynamics

The growing tendency of uncontrolled spontaneous movement

Until 1960, natural forest covered almost the entire area of the Central Highlands. During this time, the region was only very sparsely settled by indigenous groups. At that time, the population was less than half a million. This figure, however, increased to one million in 1976, 3 million in 1996, and 4.2 million in 2002. This explosive population growth is only partly explained by natural population growth, which is 2.7 percent per year in this region. An even more important contribution has been immigration.

Data provided by provincial authorities show that spontaneous immigration to the East, South, and Central Highlands has vastly exceed the rates that were envisioned in state planning. For example, Dong Nai, a province nearby Ho Chi Minh City, received more than 450,000 in-migrants between 1981 and 1989. Only 8 percent of these were the result of government resettlement efforts. The remaining 92 percent were spontaneous migrants (T.H. Nguyen 1998). Recent studies also show a growing trend toward spontaneous migration. Table 10.2 shows the population dynamics and migration pattern in Daklak province between 1976 and 2000. The result of this immigration has been a steep increase in population densities. Population densities per km^2 have increased from only 17 in 1975 to 61 in 1995, and 94 in 2000. This process has been occurring simultaneously in other provinces.

Table 10.2. Contribution of migration to population growth in Daklak province

Period	Total population	Planned immigration	Spontaneous immigration
1976–1980	128,690	114,267	14,693
1981–1985	140,701	100,002	40,699
1986–1990	171,494	79,836	91,658
1991–1995	180,561	14,334	116,227
1997–1999	7,984	-	7,984
first half of 2000	3,090	-	3,090

Source: Statistical data of Daklak People's Committee (2000).

The causes of migration

Several surveys have demonstrated that the main reasons behind spontaneous migration into forested areas are economic (T.H. Nguyen 1998). In the specific examples that I discuss in this chapter, there are three types of factors that explain migration. Each of these factors on its own, or a combination of factors, may push or pull different social groups towards the forest frontier of regions such as the Central Highlands:

1. Constraints in the places from where migrants originate. These include: (i) Lack of agricultural land; (ii) limitation of other natural resources; (iii) inadequate economic alternatives, especially employment; (iv) related poverty and famine; and (v) population pressure.
2. Attractive conditions in the settlement locations in terms of: (i) available natural resources; (ii) availability of forestland and fertile soils; (iii) low population density; and (iv) opportunity for cash-crop plantations.
3. Socio-political circumstances, such as: (i) weakness of management measures; (ii) variation in crop prices;[2] (iii) land tenure structure; and (iv) possibilities of improvements in infrastructure and services.

These circumstances usually result in a complex pattern of in- and out-migration from one region to another. In this chapter I focus on several key issues that explain the migration patterns. First, most of the immigrants into the Central Highlands were and are underprivileged people that migrated as refugees from hunger and famine. They settled in the region as the only alternative for survival. Other groups, however, were mainly attracted by the interest in valuable resources in the locations where they migrated to, such as timber, mineral reserves, cheap land, and the opportunity for capital-intensive cash-crop plantations. It cannot be excluded that some rich people living in cities came to the Central Highlands to purchase land to grow coffee in intensive plantations. For instance, one study in So Pay commune (K'Bang district, Gialai province) noticed that 20 percent of the immigrants stayed temporarily in the forests. A portion of them tried to illegally cut timber. Others purchased land for cash crop production for some years, after which they sold the land they had occupied, and moved away (V.C. Tran 1998).

Some marked correlations can be observed between immigrant origin, ethnic identity, and type of occupation. The Kinh immigrants who moved into the Central Highlands came from the provinces of the central coast such as Ha Tinh, Quang Binh, Quang Nam, and Quang Ngai. They primarily settled in urban, semi-urban, or low elevation rural areas. Conversely, migrants that came from provinces in the northern mountainous region such as Caobang, Langson, Yenbai, and Sonla, belonged to ethnic minority groups like Tay, Nung, Dao, and M'Nong. They took to the exploitation of forest resources and practiced customary shifting cultivation. They settled in the upper elevation areas suitable for upland farming. Many of these migrants came from regions where land for customary subsistence production had declined as a result of erosion, declining soil fertility, and population pressure. The Central Highlands with their rich forest resources and fertile soils were most appropriate for their customary lifestyles.

The prices of some agricultural products had an important affect on the scale, direction, and nature of migration flows. This, indeed, explained the peak of movements to Daklak in the period between 1991 and 1995, as shown in Table 10.2. This period coincides with the cultivation "fever" of coffee, rubber, sugarcane, and pepper, all of which had relatively high market prices. Since 1997, the land available for the cultivation of those crops has become limited. This has propelled the flow of migrants towards the two Central Highland provinces, Gialai and Kontum, which still have important areas of forest frontier.

Low income and lack or inadequacy of services

As shown above, the main purpose of most migrants moving to the Central Highlands is to look for land and employment. Planned migrants have been resettled in the new economic zones to provide labor for state forest enterprises, state farms, or other projects. Resettlement projects of lowland farmers fall under Programs 327 and 773 concerning policies on the use of bare land, denuded hills, alluvial flats, and water bodies (known in short as the "re-greening program") with very limited financing to support these households. Spontaneous migrants, however, try all strategies available to them to obtain land for agricultural production. Because of their limited capital assets, many of them have turned to illegal forest clearing.

Surveys have identified the relative contributions of different mechanisms to obtain land in the Central Highlands, the results of which are presented in Table 10.3 (T.H. Nguyen 1998). On average, one household of self-settled

Table 10.3. Mechanisms of land acquisition among migrants in Daklak province

Mechanism of land acquisition	Relative contribution (%)
Allocated by government	4
Purchased	47
Self settled (clearing)	46
Others	3

Source: T.H. Nguyen (1998).

spontaneous migrants clears some 4,000 to 5,000 m^2 of forested land. On this land they practice shifting cultivation primarily to meet their daily subsistence needs. These migrants face a lack of other economic opportunities and their income remains very low. They live in conditions where there is no infrastructure and services, particularly those related to water, sanitation, health, education, communication, transportation, and energy.

Impacts caused by migration dynamics

The great diversity in scale and nature of migration has resulted in many changes in the social, economic, and political aspects of the Central Highland region. They also have had a profound impact on the communities that lived there before, and on the forest cover of the region.

Change in population structure and the erosion of customary rights

In 1921, the Central Highlands had a population of 98,000 inhabitants belonging to fifteen indigenous ethnic groups. At that time only five French nationals and twenty Kinh people lived there. In 1945, indigenous people still formed 95 percent of the population, while the remaining 5 percent were Kinh and French nationals. This population structure remained roughly constant until 1975. In 1979, however, the number of ethnic groups had increased to thirty-one, and at present this number has increased to forty-seven. Currently, ethnic minorities account for only 20 percent of the population.

The transformations described above have had an enormous impact on community structures and their role in and contribution to forest management. Among the original indigenous groups, a largely patriarchal system gradually replaced the original matriarchal system. Other aspects of traditional culture have eroded gradually and these changes have affected basic community and organization structures. The state-imposed administration has profoundly affected the self-regulating mechanism that was based on customary laws.

National policies related to ethnic communities and the associated legal framework were centered on the overall control of the country's natural resources by the state with no or little recognition of the rights or possible role of local populations in forest management.

In the customary societies of highland ethnic groups, community forests were clearly defined by village boundaries. These forests were managed by a self-rule mechanism based on highly effective village laws and regulations. With changes in these societies and the penetration of official forms of administration and organization, together with the legal barriers that limited customary resource management, the role of the traditional communities and their customary laws have decreased. Because the village communities no longer see forests as their own property, they have lost interests in controlling the access to this resource. Further, the official state forest service considered activities undertaken by local communities in the forest to be illegal. Traditional upland farming has been limited or even banned by official state laws. But rather than having the expected decline in forest encroachment, this has merely reduced the effectiveness of community customary laws. People's customary practices, however, cannot be stopped in a day or two. Upland farming has become illegal, but local people have little choice than to continue to meet subsistence needs. Because of the decreasing role of customary regulatory mechanisms, encroachment into the forests has increased.

Immigration and land tenure

Migration patterns and their impacts on wellbeing and forest cover are intricately linked with land tenure issues. New migrants arriving in the central highlands have affected existing land tenure patterns. Land tenure is entirely custom-based, as there are no officially recognized cadastral maps. As a result, there are no mechanisms to allocate land to people. At the same time, local communities have little interest in pursuing official recognition of land held under customary property. Under customary law and agricultural technology, the pre-dominant land-use practices were relatively sustainable, providing that low population densities remained. State efforts to control traditional cultivation have not only eroded the effectiveness of the village law system, but also have not provided for new land allocation. This has contributed to inequality in land ownership among communities.

The demand by immigrants for both arable and residential land has led to illegal land exchanges, from both state and customary perspectives. Local people often do not know the land-use intentions of outsiders who come looking for land in their village territories. As a result, poor farmers often rent or sell part of their land to outsiders, and then they themselves move deeper into the forest to occupy new land (see Acheampong, Chapter 6). According to V.T. Nguyen (1998), from 1994 to 1997 there were 4,876 indigenous households in the Central Highlands that sold their land and moved into the forest. This is a practice that continues today as immigrants increasingly move to more remote communities to buy land, pushing indigenous farmers to even more remote and elevated areas.

The exceptions to this pattern are immigrants who originate from northern mountain provinces and only have their labor and extensive experience in forest clearing as investment resources. Regarding their habitual approach to cultivation, these people tend to occupy unclaimed land and clear forest to start agricultural production.

Forest conversion and conflicts

From 1976 to 1990, the loss of forest area in the Central Highlands rose to 30.4 thousand ha per year. Although this rate declined in the subsequent five years, it still remained at 25.2 thousand ha per year between 1991 and 1995 (FIPI 1996). One province with especially high deforestation rates was Gialai that lost 176,268 ha of forest between 1987 and 1997. Shifting cultivation caused around 48 percent or some 84,609 ha of this deforestation. A second important contributing factor, however, is illegal logging and the logging of state enterprises without the use of proper equipment and techniques. Three additional causes of forest conversion are land clearing by inter-province migrants to grow cash crops, land clearing for new economic zones, and forest fires caused mostly by slash and burn activities.

As mentioned before, migration flows affected this forest-cover change pattern either directly or indirectly. According to a case study in 1999 some 839 households with 2,509 persons migrated to the K'Bang district. They caused a 477 times increase in forest being slashed. The result was 172 ha of natural forest being lost. In Duc Co district, in 1997 alone, 100 ha of natural forest were lost to rubber and coffee plantations (V.C. Tran 2000). Table 10.4 shows forest loss in Daklak province due to immigration.

Local governments in the Central Highlands cannot control these immigration trends. Some of the resulting deforestation has led to severe competition and sometimes conflicts between different local groups. For example, there are conflicts between forest protection authorities and local people in terms of forest resource utilization and management, or between local people and newcomers in terms of land-use purposes. Meanwhile, existing policy mechanisms and frameworks, such as power sharing arrangements between the government and local communities, existing organizational structures such as state forest enterprises, co-ops, and

Table 10.4. Area deforested between 1991 and 1997 in Daklak as a result of immigration

Year	Area deforested (ha)
1991	5,600
1992	1,432
1993	979
1994	3,026
1995	3,230
1996	No data
1997	3,277

Source: T.H. Nguyen (1998).

community groups, existing knowledge and methods of forest utilization, and benefit distribution policies are all inadequate or unsuitable for the current situation. Consequently, the forest resource base of the region continues to deteriorate as a result of the increase in population pressure and illegal activities.

Measures towards sustainable forest management under migration scenarios

Since the 1990s the government of Vietnam has proposed and enacted many policies and programs that are geared towards or give high priority to mountainous regions and ethnic minorities. Several programs have tried to promote sedentarization and fixed cultivation objectives, as mentioned earlier. These programs, it was assumed, could offer the right technologies and additional support to reduce unsustainable resource use practices, poverty, isolation, environmental pollution, and insecure land tenure. These programs would contribute to improving the economic links of rural settlements with the mainstream economic, social, and cultural life of Vietnam. This, it was expected, would assure sustained economic wellbeing, a safe environment, and a reduced pressure on natural resources.

Once implemented, however, the results of the above mentioned projects were not as good as expected; poverty and deforestation persist. The major reasons for this failure are: over-centralization of decision making on essential project components; lack of inter-agency coordination for implementation of various project activities; inactive or complete absence of farmer participation; unrealistic assumptions about the sustainability of the project after the initial inputs, resulting in emigration or neglect of the maintenance of physical infrastructure; and lack of political will to tackle key problems such as land tenure.

Additional provision that those and similar kinds of projects will have to address to become successful are the needs of immigrants that settle while the projects are going on. Furthermore, these projects need to take adequate measurements to discourage new migrants from moving into the region.

Without doubt, the conservation of tropical forests is of great importance for mankind. Not only do tropical forests have important functions in terms of climate and biodiversity, they are also an invaluably rich source of biological compounds that may serve the human community in the future. At the same time it should be kept in mind that by far the largest portion of forest clearing is made for subsistence food production. This is true not only for the local people that rely on shifting cultivation but also for an ever-increasing number of immigrants opening new land in forested areas. Efforts to conserve forest, therefore, need to address and give high priority to removing the underlying causes of hunger and poverty. If these root causes of deforestation are not addressed, conservation efforts are bound to fail in the long run.

The best measure to prevent people from moving into the forested areas of the Central Highlands is to improve their economic prospects. They need to be provided with employment opportunities, infrastructure, and social services in their homelands. To resolve the existing problems of migrants that have already moved into the region, it may be necessary to introduce short-term measures to prevent people from moving further into forests. In some exceptional cases it may be necessary to remove some people by force. This applies, for instance, to situations of encroachment in protected forests that provide important functions such as water supply. In the long run, however, it is necessary to develop strategies that combine macro-level interests with direct benefits for the people and communities living in and around the various forests.

In addition, measures can be taken that directly address the problems of unsustainable forest utilization. This may take the form of the sustainable use of the natural forest itself to capture direct benefits by local forest dwellers and maintain or enhance people's livelihood opportunities. The second strategy should include achieving sustainability of land already under food production, combined with increased production of wood outside the forests to reduce pressure on natural forests. This almost certainly will imply an increase in the proportion of cash crops, enhancing farmers' incomes and thus gradually improving living standards. If this can succeed on a large scale the pressure on the remaining forests can be removed.

Many projects in other parts of the world have followed a strategy of increasing irrigation, using fertilizers, and introducing mechanization and higher-yielding crop varieties. These measures have without doubt been successful in many cases where agro-ecological and socioeconomic conditions were favorable. In the Central Highlands, however, with its infertile soils, steep slopes, and related high risk of erosion, new technologies for improved food production are normally not applicable, either for ecological or for socioeconomic reasons. Especially in the case of protected areas, there appears to be few alternatives but to involve the local people, including the new arrivals, in forest management. This should include strengthening local capacities to conduct negotiations, manage conflicts over land ownership, and allow villagers to demonstrate their genuine interest in protecting their forests.

In order to achieve these goals I propose five closely related methodical interventions geared towards a collective learning process:

1. Participatory development of interventions: This involves engaging people in developing and promoting technical, social, and economic innovations. These innovations include resource management technologies, but also developing new regulations that can resolve specific problems. The ultimate goal is to create new possibilities for income.
2. Dealing with conflicts: This includes learning to analyze and balance the interests of different and competing actors, particularly where they relate to opposing conservation and development interests.
3. Resource use planning agreements and rules: These are geared towards mutual agreement and shared regulations that will enable the sustainable use of natural resources by different actors.
4. Village organization and leadership: This is the next step, as an adequate village organization is necessary to implement agreements and rules. Crucial to this process is the strengthening of village leadership. Strengthening village organization and village leadership will also improve representation and the negotiation positions of the villages vis-à-vis, for instance, forest protection authorities.
5. Resource conservation advice: Technical knowledge needs to be made available to rural villagers. This will improve people's awareness of certain key ecological processes, and will also draw attention to the negative impacts of certain land-use practices, or certain uses of natural resources. It should also include pointing out possible alternatives.

These five methodical innovations are closely linked, and they need to be applied in conjunction. No single one of the five innovations can be removed from this list without affecting the four others. However, these innovations alone will not lead to the desired effect. They need to be complemented with infrastructure development, institutional development, the adaptation of policy framework, and the providing of other social services.

Conclusions

There is a great diversity in the scale and nature of migration and its impacts on forests and forest use in Vietnam in general, including in the Central Highlands. The causes and effects of migration to places like the Central Highlands must be seen from different angles, from different temporal and spatial perspectives, before we can design strategies to tackle this issue. When aiming to reduce forest depletion as a consequence of migration, the following recommendations particularly apply to the highlands of Vietnam.

- Where possible, there should be a clearly defined role and responsibility in forest management for all actors that have an interest. The coordination between actors should be strengthened.
- Further research should be conducted to gain more knowledge of customary forest management and community tenure systems to have more alternatives available for tenure arrangements and communal involvement in forest protection through active management.
- A strategy of forest protection should not be based solely on making laws forbidding the slashing of forests, or strengthening the monitoring and sanctioning of perpetrators. It should also be combined with enhancing local forest use as a result of which local people will become more concerned with forest protection to safeguard their own benefits.
- It also should always be kept in mind that most deforestation in the Central Highlands is caused by people, either local people or immigrants, who are trying, often in desperation, to satisfy their basic survival needs. Therefore, legal means that try to achieve protection are bound to fail if they do not give high priority to providing sustainable livelihoods for those people.
- Forest policies and official programs should strengthen community participation in forest management and increase direct benefits from forest production activities.

Notes

1. Some of the relevant policies are: Resolution 22-TW—27/11/1989 and the Decision 72/HDBT—15/3/1990 on promoting the socioeconomic development of the mountainous areas; Decree 660/TTg—17/10/1995 dealing with migration to the Central Highlands; Decree 393/TTg—

10/6/1996 on settling populations, improving infrastructure conditions, and restructuring production in the mountainous areas; and Decree 656/TTg on investment and the socioeconomic development of the Central Highlands.

2. The variations in crop prices may result in cultivation fevers that in turn affect migration flows.

References

Department of Forest Development (DFD). 2001. *The National Five Million Hectare Reforestation Programme (1998–2010)*. Hanoi: Agriculture Publishing.

Forest Inventory and Planning Institute (FIPI). 1996. Data on forest resources, 1996.

Nguyen, T.H. 1998. Spontaneous migration of people from northern provinces to the south and the problems of forest protection. *Vietnam Forestry Review*, 5.

Nguyen, V.T. 1998. Summary report of the implementation results of the inventory and assessment of the socioeconomic development impacts on the indigenous people's lives in the central highlands through the renovation years. Unpublished report.

Nguyen, X.N. 2001. Experiences and future directions of forestry funding in Vietnam. International workshop on financing sustainable forest management. In M.E. Chipeta and M. Joshi eds., *Financing Sustainable Forest Management: Proceedings of an International Workshop*, Oslo, Norway, 22–25 January 2001 Bogor, Indonesia, CIFOR. 1 CD-ROM.

Statistical data of Daklak People's Committee, 2000. DakLak Statistical Department.

Thai, V.T. 1972. *The Forest Vegetation of Vietnam*. Hanoi: Technical Publishing House.

Tran, N.P. 1970. *Preliminary Research on Natural Forest of the Northern Vietnam*. Hanoi: Technical Publishing House.

Tran, V.C. 1998. Study on social forestry in K'Bang district. Unpublished report.

Tran, V.C. 2000. Research into peculiar features of the swidden system and the participation of the Bahnar people in the use, management and protection of the natural forests in Gialai province (the case of Sopay and Solang communes, K'Bang district). Unpublished report.

11
Movements, Social Entitlements, and Economic Fortunes in the Forests of the Interior of Borneo

Cristina Eghenter

Introduction

At a time when the linkages between indigenous people, sustainable forest management, and autonomy are intensely debated and part of the agenda of political and social reforms in many Southeast Asian countries, consideration of the complex interplay of social and environmental factors becomes essential in order to avoid simplified representations of reality and seek easy, yet ineffective, solutions. Case studies can be used as analytical tools for exploring and problematizing phenomena like migration and natural resource exploitation. By highlighting the circumstances of the actors involved and by showing the connections between modes of migration, events of exploitation, and social interactions, case studies allow us to see these phenomena as negotiated processes that contain contradictions and ambiguities that, in turn, give rise to new challenges and generate the possibility for strong contestation in the communities.

In this chapter I look at one particular case of movement out of and back into the interior of Borneo in connection with the exploitation of forest products, particularly aloeswood or gaharu. I briefly review past practices of migration in the area to provide a background to recent movements of population. Subsequently, I focus on the gaharu rush and the arrival of outsiders to highlight the different circumstances of this phenomenon and explore issues of access, entitlements, and rights regarding the exploitation of natural resources. As a case study, the example I discuss is limited in time and space. Nevertheless, it can help illuminate patterns in the causal dynamics of migration and the social ecology of non-timber forest products and common resources in the interior of Borneo and elsewhere.

A history of migration in the Apo Kayan

The area examined is the highlands of Apo Kayan, an isolated mountainous plateau, or *apo* in the local language, in the interior of Indonesian Borneo. It comprises the vast area at the headwaters of the Kayan River along the border with Sarawak, Malaysia. The area is presently divided into three administrative sub-districts: Kayan Hulu with eight villages and a population of about 3,800; Kayan Hilir, with only five villages and an approximate total of 1,025 inhabitants; and the newly created sub-district, Sungai Boh, with four villages and a population of about 1,150. People are for the most part indigenous Kenyah, Kayan, and Punan. There are also a few inhabitants originating from other islands in Indonesia who came as government officials or military personnel (See Figure 11.1).

The history of the Kenyah and Kayan peoples is a history of movement. As Rousseau states, "migrations form the framework of the area's precolonial history" (1990: 23). According to several oral history sources, the Apo Kayan was first inhabited by Kayanic people such as the Ngorek (Okushima 1999). Kayan (and Modang and Ga'ai) came from Sarawak in the seventeenth century. The Kenyah entered the Apo Kayan only later, after earlier settlers had already left the area. The Apo Kayan is now considered a Kenyah "homeland". In the past, the main reasons for migrating were: inter-group warfare, leadership disputes, religious beliefs, demographic and environmental changes, or a variable combination of all these factors (Coomans 1987; Eghenter 1999; Rousseau 1990, Roodenburg 1935; Guerreiro and Sellato 1984; Whittier 1973).

In 1911, the first Dutch post was opened in the Apo Kayan, (Black 1985: 284) and the area remained under Dutch rule until 1942. The Dutch administration made definite attempts to regulate migration and fix settlement patterns in order to control the interior populations (King 1993; Sierevelt 1927). The military was especially determined to put a halt to migrations and resettlements. By contrast, colonial civil officers, or *controleurs*, often maintained a more pragmatic and favorable approach (Sierevelt 1927). Some *controleurs* were convinced that, because of the scarcity of agricultural land, part of the lower Apo Kayan population would eventually have to migrate to the lower Kayan and those on the upper part of the Apo Kayan to the Mahakam. These predictions proved correct in both cases.

There was only one documented case of incursion by outsiders into the Apo Kayan in search of forest products (Iban from Sarawak in search of gutta percha). These actions were limited spatially to parts of the forest

Figure 11.1. Map of Apo Kayan customary land

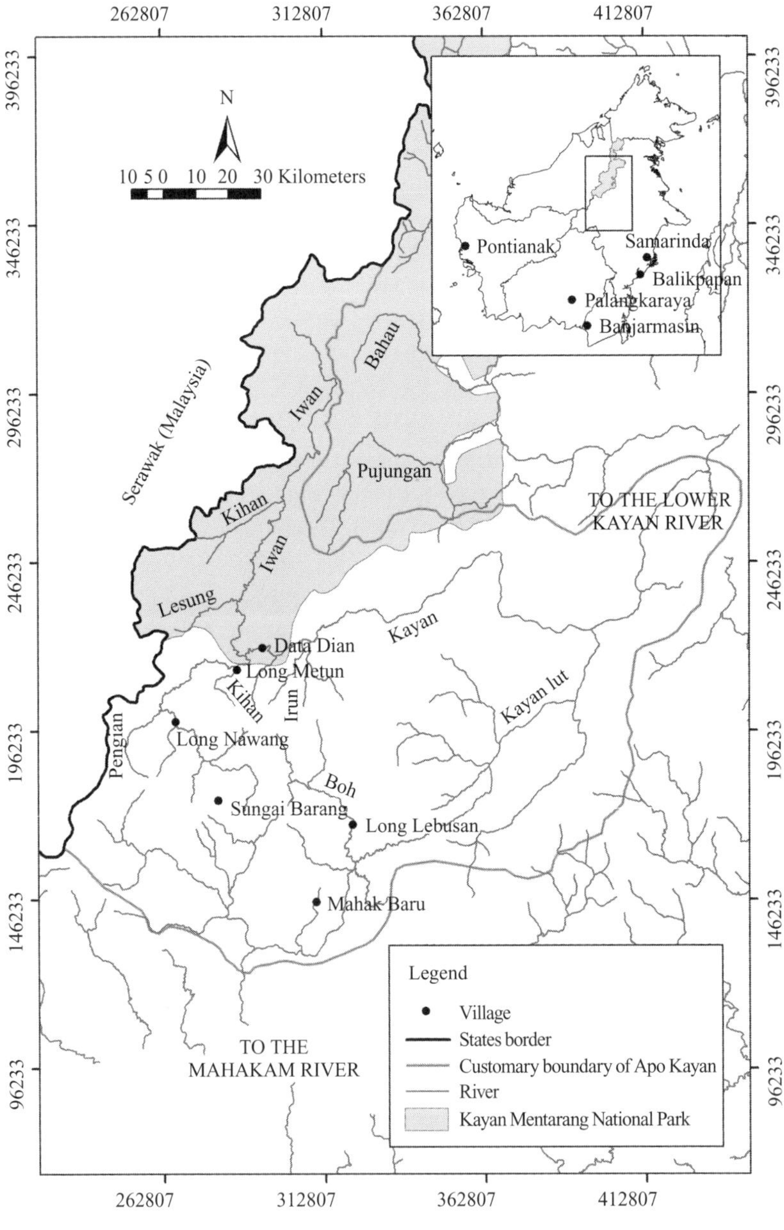

and temporally to a short period of a few years. The colonial administration directly handled the situation together with the traditional leadership in an effort to stop illegal harvesting. The end of the gutta percha rush eventually put an end to the social conflict between local inhabitants and outsiders.

The Dutch colonial administration promoted pacification, and since the late 1920s warfare has ceased to be an important reason for migrating. As a result of Dutch and American missionary activities beginning in the late 1940s, animistic religious beliefs also became less of a factor in the relocation of settlements. Religious ideals continued to have an influence, however. For example, some communities in the 1950s split into those who adhered to the Bungan cult and those who followed Christian tenets (King 1978). After Independence, and particularly during the confrontation between Indonesia and Malaysia in 1963–66[1], interior communities were officially discouraged from migrating for security reasons and also because authorities feared that the entire area at the border with Sarawak would become empty and thus vulnerable to enemy occupation.

The possibility of improving economic conditions and gaining access to the social and economic networks of the coastal urban centers attracted migrants between the 1960s and the 1980s (Eghenter 1999; Guerreiro and Sellato 1984; Dounias and Leclerc, Chapter 7). In the remote areas, there are no roads, and movement is possible only by small planes or by boating along rivers that are often disrupted by rapids. Hence it is difficult to procure in a reliable manner and at fair prices basic products such as salt, tobacco, and gasoline. Moreover, there is no convenient access to a market to sell local produce. The lack of educational opportunities for future generations together with the problem of receiving adequate health care represented two other important reasons why people abandoned the villages of the interior.

The Indonesian government continued to discourage unregulated initiatives of voluntary resettlement by interior peoples. Notwithstanding this official disapproval, no drastic measures were taken against the migrants and voluntary movements of both small groups and entire communities continued, although less frequently, into the 1980s. Some 15,000 people have moved out of the Apo Kayan since the 1960s. The majority of the people resettled in the Indonesian lowlands, but there have been occasional moves across the border into Sarawak. The last resettlement took place in 1992 when a group of Kenyah split and moved to the Upper Baluy River in Sarawak.

Characteristics of past waves of migration

My study in Apo Kayan (1991–1992, 1993) described in detail the modes and modalities of Kayan and Kenyah long-distance migrations (Eghenter 1999).

These are unique in that they typically involved a large number of people, sometimes the entire village, traveling across sparsely populated territories to a distant destination under the leadership of the village chief and/or other influential elders. The prevalent physical and social environments (harsh climatic and topographical conditions and the tight social fabric of the communities) created specific conditions under which the collective realization of the migration seemed the only and obvious way to proceed. These past migrations included both short-distance moves to other locations in the same area and long-distance migrations to a river on the other side of the watershed, or to areas as far as Sarawak, the lower Kayan River basin, Berau, or the middle and lower Mahakam River basin. According to the oldest informants, people in the past tended to move on foot rather than by boat, and to select strategic settlement sites on top of a hill or on a mountainside for defense from enemy attacks. Since the 1960s, the migrants have headed for distant locations downriver (in the lowlands), that require months or even years of travel to reach. Short-range movements have been rare with the possible exception of the relocation of temporary settlements in the lowlands.

Relocations of groups of migrants did not in general cause conflicts over land. Such movements were negotiated by the elders long before the move commenced. The prospective migrants always asked the original inhabitants for permission to relocate to their area. Permission was hardly ever denied. The agreement was also formally sanctioned in a customary way by presenting the chief in the new settlement with prestige items such as gongs, ceramic jars, and/or adorned knives. This way the group of migrants obtained legitimate rights to build the new settlement, to use and claim surrounding agricultural land, and to access and exploit resources in the forest. In contrast, those who moved out automatically relinquished rights to land and resources in their homeland, particularly if they formed an entire group or village.

The circumstances described above are indicative of "traditional" migrations and movements in the interior of Borneo. Recent economic developments have, however, led to a reversal. Now people are moving into the interior and new challenges have emerged.

The gaharu rush

In recent years, there have been no group migrations from the interior of Kalimantan to the lowlands. Hundreds of outsiders and a large number of "return migrants" have, however, come from the lowlands to the interior. This movement was triggered by the desire to seek economic fortune in the rich forests of the interior. In particular, migrants sought aloeswood or gaharu, the trade name for the fragrant resinous wood from trees of the genus *Aquilaria*[2] that have been infected by a fungus. Though recent movements do not compare with earlier migrations in terms of numbers involved, they have had a significant impact on tenure rights, ethics of access, and the sustainable management of forest products.

Gaharu is used as incense wood in the perfume industry and for medicinal purposes. It is solely for export; there is no local use for gaharu. Traders classify several categories of gaharu depending on the concentration of the infected part in the heartwood: the more uniformly infected, the more compact in color and appearance (blackish brown), and the higher the price (Blajan 1999).

In the early 1990s, the high prices fetched by gaharu caused large-scale exploitation that lasted for a few years with periodic surges in collecting. This first gaharu rush lasted from 1991 to 1995, with a peak in 1993–1994. In 1991, the price for first-quality products was already Rp 800,000 per kg (about US$200). The market value of gaharu increased steadily or, at least, remained constant, but rose dramatically right after the 1997 economic crisis hit Southeast Asian countries. By 1999, a kilogram of premium quality gaharu was fetching Rp 5,000,000 (about US$500), higher than any

Photo 11.1. A gaharu collectors' camp in the Apo Kayan

other forest product with the exception of bezoar stones.[3] Another surge in activities with the involvement of outsiders reoccurred in Apo Kayan and other parts of the interior of Kalimantan in 1998, and subsided again by the end of 2001.

Villagers in Apo Kayan witnessed the arrival of an increasing number of outside collectors organized in teams ranging in numbers from a few people to over sixty. The collectors were initially concentrated in a few villages upstream and, subsequently, when resources were depleted or collecting efforts required too much travel, they moved downriver into the Kayan Hilir sub-district, a vast territory for the most part uninhabited as a result of out-migration in the 1970s and 1980s.

While outsiders were coming to Apo Kayan, the trickle of return migrants also intensified, with a few hundred moving in by the late 1990s. Families, one or two at a time, were relocating back to the villages from where they had migrated twenty years earlier. They were also motivated by economic reasons and saw in the exploitation of gaharu and other forest products an opportunity to make good money. Return migrants had had disappointing experiences in the lowlands; dreams of easy wealth had not come true for many of them. Some organized their own gaharu expeditions. Others entered the "service" economy; with so many outside collectors in the area, they found work in providing transportation services and logistical support, and selling rice and food.

Modes of exploitation and local regulations

Like other areas in the interior of Indonesian Borneo, the Apo Kayan area is recognized by local Dayak people as being one customary land, or *wilayah adat*, under the leadership of a *Kepala Adat Besar* (great customary chief) who administers the customary law or *adat* with the help of the customary council or *lembaga adat.* All elected officials at the village level and prominent leaders of the community sit on a customary council. Customary law regulates the access to and exploitation of land and forest products. *Adat*, however, is neither fixed nor unchanging. At annual meetings, which usually coincide with the harvest festival, members of the customary councils meet to discuss and update regulations, and deliberate on matters of natural resource management and social issues. Modifications in traditional regulations are often necessary as a result of changing circumstances, for instance the negative effects of intensified harvesting pressure, and/or other changes in the natural environment and economic conditions.

Forest products with high use and market values such as rattan, construction timber, fish, gaharu, and other non-timber forest products (NTFPs) are commonly regulated, but not with equal degree. Gaharu, an exclusively extractive resource, appears to be only slightly regulated. It can be argued that this reflects the products' natural characteristics and distinct population dynamics (Jessup and Peluso 1986), as well as local histories of control and exploitation (Eghenter 2000). Gaharu trees are dispersed fairly evenly and at low densities over the territory (about three plants per ha), regeneration from seeds takes a very long time, and trees can be found in a variety of habitats. Moreover, only one out of thirty trees may be infected and yield tradable gaharu wood (Momberg *et al.* 1997; Soehartono 1997). Consequently, the finding of gaharu is highly unpredictable, the output of collection activities more uncertain and, accordingly, the risk of depletion relatively low. Further, gaharu does not have a history as a trade product locally and only recently acquired a high market value. The following are examples of regulations for gaharu collecting currently endorsed by *adat* in the Apo Kayan customary territory:

- Outside collectors must ask for the permission of the village chief and *Kepala Adat* to collect gaharu in the village area.
- Outsiders must pay a fee to the village treasury which goes towards village development.
- Gaharu must be collected the traditional way, in a selective manner, by leaving young trees and felling only those that are infected.

In general, local people have been felling gaharu trees selectively on the basis of external signs (color of the leaves, bark, etc.) of infection. Local people recognize that this method, while based on local ecological knowledge and experience, is not infallible. Moreover, there is no way to tell what part of the tree might contain the infected heartwood. Sometimes, gaharu collectors might decide to cut a tree with no apparent sign of infection just to check if it has any (Devung 1996). Local collectors organize small parties and typically spend between one to two weeks in the forest, mostly at times when they are not engaged in agricultural activities. This seems to have prevented the overexploitation of forest resources.

By contrast, outside collectors show a pattern of "exploit-and-switch behavior", typical of many forest product traders. They generally aim at rapid exploitation of the resource and maximum profit, moving to other areas (sometimes to other islands) generally before the resource is even depleted. As outsiders belonging to different ethnic and religious groups, these collectors tend to overlook, if not challenge, local customary law. They tend to cut indiscriminately infected and non-infected trees. Outsiders are organized in large groups and spend extended periods of time in the forest where they establish semi-permanent camps. Provisions and supplies are flown in by plane to a nearby airstrip, and then shipped by boat to the camps.

Modalities of social interactions

It is widely recognized that forest products of high economic value attract additional harvesters, especially outsiders, and draw the interest of powerful economic entrepreneurs (Dove 1993). This has also happened in the case of gaharu. Specialized gaharu collecting activities in the Apo Kayan were conducted and controlled by outside collectors and sponsored by Chinese and Arab traders based in lowland towns. Most of the outside collectors were transmigrants from the islands of Java, Lombok, and South Kalimantan, and hardly included any local villagers. The groups were often led by middlemen, themselves outsiders. The latter were in charge of overseeing operations, providing logistical support to the collecting teams in the forest, and storing the collected gaharu.

Upon their arrival in the villages, middlemen tried to establish amicable relationships with the villagers, particularly governmental and traditional authorities. It was essential that they build a social network to gain the necessary trust and secure access to easy "credit" from local people. This credit in the form of goods, such as rice, was to be paid once money from gaharu collecting was received. The middlemen used several ways to gain the trust of villagers. They tried to become close to local decision-makers (customary chiefs) by providing small gifts (for example, garlic or gasoline). They would promise help and give donations to the villages (such as a rice milling machine), or they would marry into a local family.

These modalities of social interaction worked effectively for the outsiders as they created dependency and bonds of moral reciprocity between them and particular families in the villages. However, in only a few cases did efforts to gain "insider" social status work on a permanent basis. Other economic and social obligations continued to tie the middlemen to the traders in the lowlands who had sponsored them in the first place. This became particularly evident when events in the village seemed to threaten the strength of the "social contract" between the middlemen and the villagers. If the villagers decided to collect on old debts or demand larger donations in return for the right to continue to exploit gaharu in the area, the middlemen often chose to leave and return to the lowlands. Middlemen also established privileged relationships with local shopkeepers and government employees. The latter, often individuals originating from different villages or different ethnic groups, eventually would become local trusted "assistants", or start their own career as "middlemen".

At a community level, the strategies of the middlemen created division and strife. The initial differences of opinion between those supporting outside collectors and those challenging the legitimacy of their presence grew into a rift as a result of the middlemen's attempts to win the support of a particular

Photo 11.2. Gaharu collectors who work for a local entrepreneur resting from their work in a collectors' camp

family or kin group in the community. Those who felt excluded from the new social connections and economic advantages of these associations had additional reasons not to support outsiders.

Outside gaharu collectors were at first welcomed in the Apo Kayan area. They came amid many promises of economic assistance at a time when gaharu collecting was not a common activity among local residents. Initially, outside collectors were not asked to pay collecting fees. It can be argued that the presence of traders encouraged more residents to try to undertake expeditions to the forest in search of gaharu or other forest products like bezoar stones. These enterprises were, however, limited in scope and time, and distinct from the large-scale operations directly sponsored by the traders. However, as the level of exploitation and the number of outside collectors increased, were any management actions or controlling measures taken by local residents? Were any efforts made to relieve social divisions in the community? Questions of this kind touch on definitions and practices of social space. They also relate to issues of boundary keeping and the ethics of access (Peluso 1996).

Definitions of social space and the nature of boundaries

In Kenyah communities, individual claims to land are established by cutting trees or clearing forest. The right to use the land is then passed on to the successive generation. Useful trees such as fruit trees, cinnamon, and bee trees (*Kompassia spp.*) are owned by individuals or kin groups (Peluso 1996). Traditionally, when people decided to move to another community, they would transfer their rights to fallow land and trees to family members or, in more recent times, sell the rights to other villagers. This would void future claims to land or resources unless family members continued to live in the village.

Another form of land and forest tenure is locally known as *tana ulen*. *Tana ulen* refers to land with primary forest rich in natural resources that have a high use-value for the local community, and where the use of forest products is restricted. In the past *tana ulen* functioned as forest reserves managed by the aristocratic families of the community. Nowadays, responsibilities for the management of the forest reserves have been transferred to the customary councils that oversee *tana ulen* forests on behalf of the entire community and according to customary law (Eghenter 2000). All other forests in the village territory may be regarded as community land with open access regulated by social norms. The communities know the boundaries of the territory, but these are not enforced in an exclusive way. The lack of boundary demarcation helps retain an element of fuzziness for outer boundaries. The latter aspect is clearly present in the cognition and use of territory, and the resources within, by mobile peoples (Casimir 1992).

Casimir (1992: 10–13) distinguishes between two strategies in the enforcement of territory: "spatial boundary defense" (perimeter defense), and "social boundary defense". The first strategy occurs where resources are dense and predictable. The sense of territoriality grows weaker the farther away from the center. In contrast, in a "social boundary defense" strategy, resources are usually unpredictable and scarce, and access is regulated more by the social group than by means of a marked territory.

The "social boundary defense" strategy can be applied to the circumstances of the people living in the Apo Kayan. There is a general sense of "the territory", the Apo Kayan customary land, but this does not mean that the area is rigidly bounded along clear-cut lines. Neighboring people can access the area provided that they ask the *kepala adat* for permission. The use of the territory can always be negotiated between neighboring groups. Boundaries are enforced more like open and flexible social spaces. They are regulated by the considerations and needs of the social group more so than by the rigid interpretation of a divided physical space. When chasing wild pigs, for example, hunters from one community often cross the boundary into the territory of another community until they succeed in catching their prey.

Who is entitled to gaharu?

A cognitive view of social space as described above imposes a certain degree of flexibility in the enforcement of boundaries and boundary rules. Wadley (2002) argues that three main criteria enter in the negotiations for requesting and granting access to resources: entitlements (who has the legal right to exploit resources?); identity (shared ethnic, cultural, and religious characteristics); and reciprocity (obligations and expectations). The process of negotiation exhibits, however, many ambiguities and contradictions, and the outcome of the process cannot be predicted solely on the basis of the extent to which the three criteria are fulfilled. Decisions to restrict access reflect a very discretionary interpretation of the social space. Social and political circumstances, economic favors and rewards affect boundary interpretation and ethics of access. Regulations might state a matter of principle (the exclusion of a user group; guidelines for harvesting specific resources), but do not rule out the possibility for modifications, adjustments, and compromise. The more flexible the rules, the more room a boundary keeper may have to negotiate conditions of access, and the more chance outsiders may have to be granted access to a particular resource (Wadley 2002). It must also be added that, as in the case of the gaharu rush in Apo Kayan, the stronger the "social contract" between collectors and local families, and the deeper the web of obligations, the more chances an outsider may have to continue to enjoy unlimited access to the community resources.

In response to the increasing number of outside traders, customary councils in the Apo Kayan deliberated on the need to stop outside gaharu collectors from accessing their land. As a measure they proposed to confiscate the traders supplies and belongings. *Adat* leaders were very concerned about the situation, but sometimes lacked the necessary legal authority and internal consensus to impose their will. In some villages, the village administration started to impose user fees, but for the most part these were minimal. They did not limit the number of outside collectors, but increased by a small share the benefits of local residents. It was more practical to exact a fee from users than it was to impose a tax or a percentage fee on the amount of gaharu collected. Collectors did not report the amounts harvested and gaharu could be easily smuggled out of the forest without passing through villages. In one village in the Apo Kayan, Data Dian, for example, the user fee was set at Rp 30,000 per collector for each month spent in the forest in 1999.

Attempts to exact the fees were often unsuccessful. On the one hand, if the locals pressed the case too much, outsiders began challenging the legitimacy of the decision by arguing that local communities had no right to impose the

Photo 11.3. Construction of a temporary house in a gaharu collectors' camp

fees as this was state forest (belonging to the Indonesian state) to be used and managed for the welfare of all Indonesian people. On the other hand, locals themselves balked at taking overly drastic measures against collectors such as the confiscation of gaharu or personal belongings. In the customary meetings, proposals were made to raise fees rather than stop collectors. In part, the rhetoric revealed a genuine sense of compassion for the collectors. It also disguised the economic interests and personal obligations of the families already "indebted" to the traders by virtue of having accepted gifts and financial help. Eventually, the customary council assembly ruled to raise the collecting fees. In 2001, these fees rose to Rp 300,000 per collector per month and Rp 500,000 per trader.

In the view of the customary council, raising collecting fees was a "fair" solution. In principle, it reiterated the right of local people to restrict access and thus protect "their" forest resources; it also dismissed the existence of internal divisions in the community under the formal appearance of consensus. In reality, the enforcement of old and new regulations continued to be hindered by the conflicts of interest within the communities. Discontent among those families that were not benefiting economically from the gaharu business or the "social contract" with the middlemen was growing, as were tensions and divisions. The youth voiced strong opposition towards the decisions of the traditional leadership, accusing them of being too easily influenced by the generous tactics and words of the traders. Outside collectors were taking advantage of the existing rifts in the communities.

Enforcement requires broad social support so, as long as divisions existed, it was not possible.

Beginning in 1999, a couple of local entrepreneurs independently entered the Apo Kayan gaharu business, and increasingly employed teams of outsiders. They were ostracized by the Chinese traders in the coastal towns and had difficulty selling gaharu for good prices. Other residents of the Apo Kayan were also getting into the gaharu network by sponsoring a few collectors. They were, however, acting more like middlemen and reported to one of the big Chinese or Arab traders. Acting out of frustration, the Apo Kayan entrepreneurs decided that it was time to write a letter to the provincial government to ask that all outsiders be ordered to leave the Apo Kayan. They wanted, however, the legal approval and social support of the customary council. In January 2001, the customary council of Apo Kayan issued a written, formal statement asking gaharu collectors still operating in the Apo Kayan to leave the area ("No traders of gaharu, no collectors nor collaborators are allowed to carry out their activities in the Apo Kayan area anymore"). After the issuance of the decree, the customary council planned a survey to the Iwan River, one of the catchments in the Apo Kayan where gaharu was collected, to check whether collectors were leaving the area. The survey took place in April of the same year. Subsequently, very little enforcement was imposed in the following months. Fines or outstanding debts were rarely collected. Apart from the physical difficulties of controlling access to the forest and monitoring the vast area, internal divisions, bribes, and conflicts of authority weakened *adat* and undermined its effective role in enforcement.

Towards the end of 2001, most gaharu collectors had left the Iwan River. Some had left the Apo Kayan to go back to the coastal towns; some had trekked across the watershed to the Pujungan River, hoping to seek fortune in that area; still others moved downriver. It is doubtful that this outcome was solely the result of the customary council's actions. Economic considerations had weighed more on the decisions of traders and collectors. Indeed, recently, collectors were only using the base-camps on the Iwan River to launch their gaharu expeditions across the border into Malaysia.

Contested entitlements and poor management of forest products

The literature on common pool resources describes conditions that need to be met before sustainable common pool resource management is possible: clear boundaries; clear membership criteria with rules for inclusion and exclusion;

the capacity to monitor behavior; equitable distribution of power and resources; and well-defined property rights (see for example, Ostrom 1999; Tucker 1999; Gibson *et al.* 2000; Wollenberg *et al.* 2001). In the Apo Kayan, the criteria for membership to the groups claiming common pool resources became a challenge. It was relatively easy to categorize outsiders with no local relations as "non-members". This issue became fuzzier, however, in the case of return migrants or outsiders who had married into the community. People, by virtue of belonging to the same ethnic and family group, might be "insiders" and as such may legally claim a share of the resources and land in the village of origin. However, according to the *adat* rules, they had relinquished all claims as a result of having moved out of the area, and thus are considered to be "outsiders". Their social position is ambiguous; they are in the in-between position of "outside insiders".

No unanimous policy was proposed on this issue. While in some villages local people have allowed return migrants membership as "insiders", in other villages local people have taken an uncompromising position and imposed the same restriction rules on former residents as on outsiders. In the first case, local people have recognized the negative social consequences of an act of exclusion (the foregone reciprocity or severance of the possibility of social expectations and obligations) over economic loss (the opportunity cost), and thus given family links more consideration.

Perceptions of illegitimacy also changed. These in return affected the sense and definition of rights: who is entitled, and who is not, to exploit natural resources. The boundaries of the customary land were unquestioned, but the ethics of access were shifting. In the beginning, Kenyah leaders in the Apo Kayan accepted the presence of outside collectors and did not consider them a threat to gaharu. They started challenging the legitimacy of their presence and complained about the diminished returns of their own collecting activities after initial promises of help and donations went unfulfilled.

The modes of exploitation by outsiders in conjunction with social divisions have weakened the sustainability of natural resource management. In a traditional situation, customary laws combined with the seasonality of expeditions of local people prevented overexploitation. The unpredictability of a resource such as gaharu, often conceptualized in terms of *nasib* or good fortune by local people, reinforced the expectations of local collectors that the resource would not be easily depleted and that there would be no immediate need to control outsiders. However, specialized collectors were able to survey larger and more distant areas, and focus exclusively on this extractive activity. Gaharu might not have been exhausted in a biological sense, but it was reaching a level of local "economic non-viability", that is, collecting activities were getting beyond the economic means of local people. A successful expedition to the forest required a higher financial burden and more travel time for local collectors, who might not be able to afford it.

The economic situation of the Apo Kayan was also negatively affected by the "gaharu rush". It created an unusual "one trade commodity" economy with unstable market prices for all goods. The traders could afford to purchase goods and services at much higher prices than the locals thus creating an inflationary trend. From a different perspective, the donations and gifts of the traders were not comparable to the market value of the gaharu collected and traded out of the Apo Kayan. In this regard, there was no equitable distribution of resources. Also, while local people linked to the traders benefited appreciably from these activities, most people did not.

Most recently, outsiders are coming back to the Apo Kayan. People in the communities feel unable to monitor the presence or arrival of more collectors. The area is too vast and there are too many shortcuts or tributaries that enable people to enter the forest without being seen or known. Although the amount of gaharu collected in one expedition has drastically diminished (more than 50 percent of the expeditions fail), and the quality has degraded, the market prices have gone up to reach Rp 5,000,000 per kg for poor quality gaharu. The possibility to fetch high prices compensates for the higher risk of failure in an expedition. More local people also engage in the exploitation of bezoar stones as an alternative to the diminished returns of gaharu collecting. Moreover, for outside collectors in search of gaharu, the Apo Kayan is increasingly becoming the way to enter Malaysia (illegally).

Divisions in the communities, the coming of outsiders, and strategies of "imposed" social and economic obligations practiced by the traders, negatively affect the potential for cooperation in the community. A one-commodity economy and unequal economic opportunities undermine the long-term security and wellbeing of the local people. Fuzzy "membership" generates more ambiguity and contradictions in the local definitions of ethics of access, which are increasingly becoming sites of contestation rather than structural principles. Although efforts continue at a community level, equitable solutions in the management of gaharu and other natural resources appear difficult to sustain under the present social and economic circumstances.

Acknowledgments

This paper was drafted on the basis of field data and information collected during fieldwork in the interior of Kalimantan and during my tenure at WWF

Indonesia. I am grateful to Liman and other team members and friends for fruitful and engaging discussions on the topic. I hope that their commitment to trying to promote sustainable management of natural resources in Apo Kayan will continue. I also thank Lye Tuck-Po for her precious comments and suggestions on a previous draft of the paper.

Notes

1. The area at the border between East Kalimantan and Sarawak was used as the country security frontier from where (small) military incursions were conducted by the Indonesian army deep into the jungle of Sarawak in support of the *Tentara Nasional Kalimantan Utara* (National Army of North Kalimantan) (Mackie 1974).
2. In Kalimantan, there are three species of *Aquilaria*: *Aquilaria malaccensis, A. microcarpa, and A. beccariana.*
3. Bezoar stones are found in the bladder of leaf monkeys (*Presbytis spp.*) and porcupines (*Hystrix spp.*), and are used in Chinese traditional medicine.

References

Black, I. 1985. The "Lastposten": Eastern Kalimantan and the Dutch in the nineteenth and early twentieth centuries. *Journal of Southeast Asian Studies*, 16(2): 281–291.

Blajan, K. 1999. Jaringan Pemasaran Gaharu, Pengelolaan Hutan, dan Dampak Sosiologis, Ekonomis, dan Ekologisnya si Kawasan Sungai Bahau. In C. Eghenter and B. Sellato, eds., *Kebudayaan dan Pelestarian Alam. Penelitian Interdisipliner di Pedalaman Kalimantan* (The trading Network of Gaharu, Forest Management, and the Social, Economic, and Ecological Impacts in the Bahau River area), 181–200. Jakarta (Indonesia): Ford Foundation and WWF.

Casimir, M.J. 1992. The dimensions of territoriality: An introduction. In M.J. Casimir and A.Rao, eds., *Mobility and Territoriality: Social and Spatial Boundaries among Foragers, Fishers, Pastoralists and Peripatetics*, 1–26. Oxford: Berg.

Coomans, M. 1987. *Manusia Daya. Dahulu, Sekarang, Masa Depan.* (The People Upriver. Past, Present, Future). Jakarta: Gramedia.

Devung, S. 1996. *Praktik dan Pranata Tradisional Pemanfaatan dan*

Pengelolaan Sumber Daya Hutan di Daerah Sungai Bahau, Kalimantan Timur. (Practices and Regulations in Traditional Use and Management of Forest Resources in the Bahau River Area, East Kalimantan). M.A. Thesis, Program Pascasarjana Program Studi Antropologi, Universitas Indonesia.

Dove, M. 1993. A Revisionist View of Tropical Deforestation and Development. *Environmental Conservation*, 20(1): 17–24.

Eghenter, C. 1999. Migrants' practical reasonings: The social, political, and environmental determinants of long-distance migrations among the Kayan and Kenyah of the Interior of Borneo. *SOJOURN, Journal of Social Issues in Southeast Asia*, 14(1): 1–33.

Eghenter, C. 2000. What is Tana Ulen good for?: Considerations on indigenous forest management, conservation, and research in the interior of Indonesian Borneo. *Human Ecology*, 28(3): 331–357.

Gibson, C., M. McKean, and E. Ostrom. 2000. *People and Forests. Communities, Institutions, and Governance*. Cambridge: MIT Press.

Guerreiro, A. and B. Sellato. 1984. Traditional Migration in Borneo: the Kenyah Case. *Borneo Research Bulletin*, 16(1): 12–25.

Jessup, T. and N. Peluso. 1986. Minor forest products as common property resources in East Kalimantan, Indonesia. In Proceedings of the Conference, *Common Property Resource Management*, 21–26 April 1985, 505–524. Washington DC: National Academy Press.

King, V.T. 1978. Revitalization Movements in Kalimantan. *Indonesia Circle*, 17: 14–27.

King, V.T. 1993. *The Peoples of Borneo*. Oxford: Blackwell.

Mackie, J.A.C. 1974. *Konfrontasi. The Indonesia-Malaysia Dispute 1963–1966*. Kuala Lumpur: Oxford University Press.

Momberg, F., R. Puri, and T. Jessup. 1997. Extractivism and extractive reserves in the Kayan Mentarang Nature Reserve: Is *Gaharu* a sustainably manageable resource? In K. Sorensen and B. Morris, eds., *People and Plants of Kayan Mentarang*, 165–180. Jakarta: Unesco and WWF Indonesia.

Okushima, M. 1999. Wet rice cultivation and the Kayanic Peoples of East Kalimantan: Some possible factors explaining their preference for dry rice cultivation. *Borneo Research Bulletin*, 30: 74–104.

Ostrom, E. 1999. *Self-Governance and Forest Resources*. Occasional Paper No. 20. Center for International Forestry Research, Bogor, Indonesia.

Peluso, N. 1996. Fruit trees and family trees in an anthropogenic forest: Ethics of access, property zones, and environmental change in Indonesia. *Comparative Studies in Society and History*, 38: 510–548.

Roodenburg, W.P. 1935. *Memorie van Overgave van de Onderafdeling Apau Kajan*. Unpublished manuscript. (Koninklijk Instituut voor de Tropen, Amsterdam, 1075).

Rousseau, J. 1990. *Central Borneo: Ethnic Identity and Social Life in a Stratified Society*. Oxford: Clarendon Press.

Sierevelt, A.M. 1927. *Memorie van Overgave van de Onderafdeling Apau Kajan*. Unpublished manuscript. (Koninklijk Instituut voor de Tropen, Amsterdam, 1067).

Soehartono, T. 1997. The overview on Trade in *Gaharu* in Indonesia. Paper read at the 3rd Regional Workshop for Conservation and Sustainable Management of Trees, 18–21 August; Hanoi, Vietnam.

Tucker, C. 1999. Private versus common property forests: Forest conditions and tenure in a Honduran Community. *Human Ecology*, 27(2): 201–230.

Wadley, R. 2002. *Letting Outsiders In: Forest Product Collection, Boundary Keeping, and Resource Management in West Kalimantan, Indonesia*. Unpublished manuscript.

Whittier, H. 1973. *Social Organization and Symbols of Social Differentiation: An Ethnographic Study of the Kenyah Dayak of East Kalimantan (Borneo)*. PhD thesis, Michigan State University.

Wollenberg, E., A.A. Nawir, A. Uluk, and H. Pramono. 2001. *Income is Not Enough: The Effect of Economic Incentives on Forest Product Conservation*. Bogor, Indonesia: Center for International Forestry Research.

12
We Come to Grow Coconuts, but Not to Stay: Temporary Migrations into the Peat Swamp Forests of Sumatra

Abe Ken-ichi

Introduction

This chapter examines what happens to communities several generations after migrants first settle in a forest frontier; the peat swamp forests of Sumatra, Indonesia. From the existing literature, we can expect that a combination of factors influences how migrant communities will fare. These include the biophysical environment of the forest frontier, the cultural and social features of the migrants, the wider social-economic-political conditions under which the migrants operate, and the social and cultural features of the place of migration. To our knowledge, little systematic thinking has been expended on understanding the relative importance of these four factors, or how they interact with each other to shape the long-term fate of forest frontier communities. This is the goal of this chapter.

Peat swamp forest (PSF) is a kind of tropical forest that is found on peat lands. It is a particularly harsh environment, and poses unique challenges for migrants. The specific characteristics of peat render agricultural development particularly difficult. For this reason, PSF was one of the last tropical habitats to be occupied and converted for agriculture. In Indonesia the settlement of PSF started in the middle of the twentieth century, accelerated in the 1960s, and expanded drastically in the 1980s. Given its young history, migration into the Sumatran PSF, and what happened to the pioneer communities, can easily be traced.

The concern of this chapter is, therefore, the "maturing" process of the migrant societies that settled in Sumatra's PSF during the second half of the twentieth century. Data for this study is based on two field surveys. In 1993, a complete household census was carried out in two migrant settlements in Sumatra's PSF as well as a larger village that provided various services to one of the two settlements (Figure 12.1). The survey focused on social structure, family history and household economy. The larger village was revisited in 1997, and a follow-up study of the migrant communities was conducted in 2001. The remainder of this chapter presents the analysis of these surveys. The results contribute to the understanding of how the forest environment, the migrants' socio-cultural attributes, the dominant social-economic-political environment, and the socio-cultural conditions at the frontier can explain some of the outcomes of the settlement and adaptation patterns of PSF migrants.

This chapter is divided into four sections. In section two I describe the history of PSF occupation in Southeast Asia, and especially in Indonesia. In

Figure 12.1. Location of the study sites

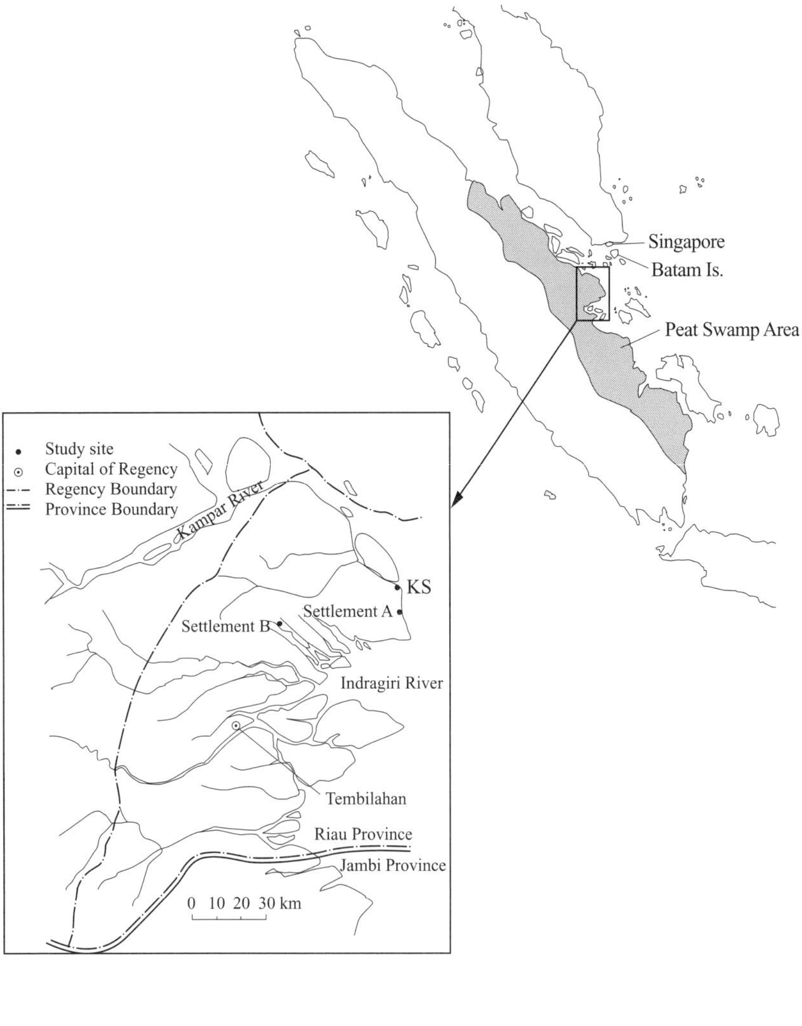

section three I detail the characteristics of the two migrant settlements and a larger village where many of the pioneers eventually moved to. In section four I look at the contemporary situation and extrapolates into the near future. In the last section I draw general insights from the results presented below.

Peat swamp forests: Setting and history

Tropical PSFs are found exclusively in insular Southeast Asia (Whitmore 1984). One of their unique ecological characteristics is forest exceeding 40 m in height that grow on peat rather than on soil. Until recently, peat was thought to be distributed only in cold climate regions, where under wet conditions plant material would accumulate rather than decompose. But in permanently humid conditions, such as in certain tropical regions that receive high precipitation throughout the year, the decomposition of plant residue is also restrained. In sum, PSF is a delicate ecosystem[1] with extraordinarily high humidity that has long resisted human attempts to transform it (Abe 1993, 1997).

The areas surrounding PSFs were not utterly devoid of inhabitants. Malay peoples settled in fishing villages along seacoasts and riverbanks throughout the Southeast Asian archipelago. They subsisted on maritime fishing (see Photo 12.1), collected *nipa* palms for roofing and construction timbers from the mangroves, but seldom ventured deep into the PSF. Others, like the Chinese loggers called *Panglong*, entered PSFs to collect forest products (Erman 1994). From the late nineteenth until the early twentieth centuries, the *Panglong* harvested PSF timbers to meet the growing demand in Singapore for construction materials. Many of the *Panglong* became exhausted, sickened, and died in this unhealthy working environment (Pastor 1929). The overall impact of *Panglong* on PSFs was small, given the simplicity of the logging tools they used. The irreversible deterioration or destruction of PSFs only began when people came into the forest to settle with the intention of converting the forest to productive lands.

Migration to the PSF of Sumatra started in the 1940's when groups of Banjarese fled from turmoil in their homelands in Kalimantan and discovered unoccupied PSF in Sumatra. The Banjarese had lived in PSF at the lower basin of the Barito River in Kalimantan and brought their knowledge with them. Converting PSF demands special techniques. When the soil that is commonly found below the peat layer is exposed to the air after the loss of the upper layer, it emits poisonous sulfuric acid. It is, however, possible to create relatively stable agricultural lands by constructing canals and locks

Photo 12.1. View of KS, a village raised on piles and projecting into the sea

through which water from rivers is channeled into the fields at high tide and sulfuric acid is rinsed out at low tide. This is a technique known in Sumatra as *pasang surut* or tidal irrigation. The Banjarese brought this technique to their new homes in Sumatra. They cleared forests, dug canals, planted rice in irrigated fields and harvested coconuts that grew on mounds in the paddy fields. The Banjarese knew that tidal irrigation is only effective where there is a significant difference between high and low tides and where there is a large river. As a result, they settled only in peripheral peat areas such as the banks of rivers. They never moved deep into the PSF.

The second wave of migration to the Sumatran coastal PSF in Riau and Jambi Provinces started in the 1980s and has continued ever since. This time, however, the migrants specifically looked for land to develop as commercial coconut plantations. Coconut oil was in high international demand and the high prices were attractive for small producers. In addition to the Banjarese, Bugis came from Sulawesi and Javanese from Java; even local Malays, who had been living along the Malacca straits, moved to PSF areas.

When the Bugis came, the peat areas suitable for rice cultivation by tidal irrigation were already occupied by earlier migrants. However, coconut plantations need no irrigation. Coconuts can be planted for a limited duration

Photo 12.2. A Batak "dentist" at a periodical fair. Notice the pile of extracted teeth that demonstrate his skill

of time on PSF land that is free from tidal influences. After clearing the forest, coconut growers dig canals before planting their trees. The canals are prepared solely for draining excess water held by the peat. Little capital or technology is needed for this kind of forest conversion.

Peat-land agriculture incorporating tidal irrigation is relatively stable, although in the long run the paddy fields can be eroded and become subject to flooding. By contrast, coconut cultivation presents more serious problems related to the physical and chemical characteristics of peat. After the loss of the tree cover and the drainage of the peat lands, the peat dries, gradually shrinks and then decomposes. The ground level subsides, causing coconut trees to be easily felled by winds, or tree roots exposed above the ground. The subsiding ground is also more vulnerable to floods. In addition, peat is originally oligotrophic and lacks nutrients that are essential for plant growth. Where tidal irrigation is possible, nutrient-rich river water supplements the missing nutrients. Without this, coconut yields sharply decrease after the nutrients contained in peat are exhausted. The decline of yields depends mainly on the depth of peat, but may start just ten to fifteen years after the coconuts are planted. Because of these limitations, coconut cultivators are obliged to abandon their plantations after just a few years. They then move on, find fresh areas of PSF, establish new coconut plantations, and leave wastelands behind them (Abe 2003).

The sociology of peat swamp forest migrant communities

In Indragiri Hilir, Sumatra, settlements in a PSF area are characteristically aligned along canals. The first pioneers enter the PSF from a river or the sea. Once they have found a suitable location, they collectively start digging a canal into the forest. Newcomers are allotted a forest plot at the far end of the canal. Thus the earliest migrants are distributed at the beginning of the canal, and the later migrants landwards. The size of the plot allotment is measured by the number of coconut trees that can be planted in a single line parallel to the canal. Houses are spaced roughly at equal intervals along the canal. Migrants are allowed to clear forests "without limit" in a direction perpendicular to the canal. As the migrant population grows, the canal stretches farther and farther into the forest.

The particular production pattern resulting from the PSF ecology has its implications for settlement patterns and related social behavior. Once the forest is cleared and coconuts are planted, little work is required. Weeding is necessary only once every three months, and is combined with harvesting. As a result, coconut cultivators do not need daily access to their plantations. The need for security and convenience are other issues influencing migrants' preference to cluster with each other, rather than to be isolated in their coconut plantations. For these reasons—social and ecological—migrants establishing PSF coconut plantations tend to move on quite soon after they have planted the trees. They join the village on stilts at the base of the canal, and from there move on to larger Malay villages that offer more social and religious amenities, such as primary schools, Islamic schools, mosques, shops and markets.

This continuous movement of PSF migrants is reflected in the census of two settlements researched for this paper, settlements A and B. Both belong administratively to the regency of Indragiri Hilir, Riau Province. One settlement mainly consisted of Buginese and the other of Banjarese migrants. The two settlements differ in some ecological, social, and economic aspects.

In the two villages, many people had emigrated from the peat forest in the eight years since the first census (Table 12.1). In settlement A, the first migrants from Sulawesi appeared as early as 1979. After commuting from a nearby village for a few years, they started to live in their own settlement and invited relatives in their hometowns to come and join them. The number of families reached thirty-eight in 1993. In 2001, nineteen families, or half of the thirty-eight families, emigrated from settlement A. Furthermore, four families moved their houses to the base of the canal within the settlements.

Among those who emigrated, four families returned to the island of Sulawesi, their home land. Two households had sold their coconut plantations and moved to a larger town on Sumatra. Many of the people who emigrated from settlement A moved to live in a traditional PSF village raised on stilts, at the base of the canal where their estate was located, or within the same sub-district. They maintained their plantations and former houses, and visited those only for harvesting and some tending purposes. They now live most of their lives in villages with a mosque and school.

Similar trends of emigration are also observed in settlement B. In this settlement, there were a few Javanese pioneers who had arrived in 1976. The Banjarese migrants who later became a majority population only joined the settlement from the mid-1980s. The number of families had expanded to fifty-four at the time of the first survey. Twelve out of these fifty-four households (22.2 percent) migrated out between 1993 and 2001. None of them returned to their home land. Five households left the peat swamp after selling their plantations, three moved to the larger town, while the heads of two other households became wage laborers at sawmills. Another five households maintain their coconut plantations, but have moved to a nearby village.

In the same period, ten new families migrated to settlement A. All but one came from other settlements in the PSF area and were Bugis. One Bugis migrant came directly from Sulawesi. In settlement B, nine new migrant families arrived: three Banjarese, five Malay from adjacent settlements, and one Javanese of Sumatran origin. Three new families were from the

Table 12.1. Movements of peat swamp forest inhabitants between 1993 and 2001

	Settlement A	**Settlement B**
Households in 1993	38	54
Emigrated out		
Moved to a village on stilts	13	5
Moved to town	2	5
Returned to homeland	4	0
Unknown	0	2
Moved within the settlement	4	5
Total emigrated out	19	12
Newly recruited	10	12
Total number of Households in 2001	29	54

Photo 12.3. Collective logging work in 1997. Logs are sold to a sawmill nearby

same settlement. The total number of households had decreased by nine in village A, while in village B the number of households remained the same as in 1993. During the same period the heads of four households out of thirty-eight died in settlement A, and nine in settlement B, illustrating the physically demanding conditions of life in peat swamp territory and the lack of medical assistance.

The evolution of villages

Because of migrant settlement and relocation patterns, a concentration of the population occurs in villages that are raised on stilts, most of which are located at the base of the canals. These are therefore the most dynamic centers of development and change in the PSF region. One such village is KS, located in the same sub-district as settlement A. In 1992 the village had ninety-eight households. The forty-eight native Malays in 1992 all subsisted on fishing. Today export-oriented prawn fishing is the most important economic activity. In 1992 twelve households were Bugis, fourteen Banjarese, and two Javanese. Nine Chinese households were brokers and rented out fishing boats. Ten households were Orang Laut (literally, people of the sea). There were three Minangkabau households, running a restaurant, general store, and a rice mill. In 1992, fifty-eight of the ninety-eight households owned fishing boats, varying in size from engine-powered boats to rowboats. Those who did not own boats worked as laborers on fishing boats. Thus, most villagers somehow depended on fishing at that time.

The population of the KS village increased rapidly with the influx of migrants from the canal communities. The number of households reached 223 in 1997, and had exceeded 400 by 2001. Many of the newcomers still owned coconut plantations that they visited periodically to harvest and occasionally tend. These people clearly moved to KS village to benefit from its services, particularly education for the children. This is also reflected in the increase of KS primary school pupils from 160 in 1994 to 289 in 2001. The school's three classrooms cannot accommodate the increased number of pupils. As the number of children gradually exceeded the primary school's capacity, privately owned *madrasahs* or Islamic schools were built. In 2001, two *madrasah ibtidaiyah* (equivalent to primary schools), and one *madrasah tsanawiya* (equivalent to a secondary school) were established. The village still does not have a state secondary school, even though it obtained formal village status in 1999.

The increase in village population attracted itinerant traders, who for many years came to trade on fixed days of the week. There were also some vendors of sweets and beverages, and dentists and barbers also opened for business. People came from neighboring settlements, some in boats, to sell vegetables and fruit. At these markets people met and exchanged information. Most merchants traveling to and from regular markets are Minangkabau from central and north Sumatra. Once villages grow and start having their own stores traveling merchants avoid them, as has happened in KS village since 2001. By that time, the village had eighteen shops, three kiosks, one coffee shop, and three restaurants.

Characteristics and prospects of the migrant community

Since their earliest arrival, migrants continued to relocate to the once unpopulated PSF. Now, the population is increasing rapidly. The population of the sub-district that includes settlement A, for example, increased from 60,797 in 1999 to 67,436 in 2000—a 10.9 percent increase in one year (interview with sub-district head). As the population grows, schools and clinics are being built. The fate of the PSF migrants varies; some households are successful in building a new life, others are not. As the main resource of new arrivals is labor, those households that lose their breadwinner due to illness or injury face poverty. No choice will be left to them but to sell their half-finished plantations and leave the PSF. Some members of a village buy coconut plantations from these unfortunate households and become owners of large estates. Most migrants, however, do neither too well nor too badly.

Many consider it an achievement if they can move to the villages that are raised on stilts so that they can benefit from the social services there while maintaining sufficient incomes. This section analyses the aspirations of the PSF migrants and what attracts them to these villages.

Religion: Social bond and last resort

Many of the migrants who come to the PSFs of Sumatra rely on kinship or birthplace relations to start their travel. People from the same birthplace or of the same ethnic background tend to form a settlement together. Frequently, however, migrants will then move on. When people from different places of origin come to the same region, it is not unusual for multiethnic communities to form in settlements along a canal. Settlement A was dominated by the Banjarese pioneers until the early 1970s. By 1993, the settlement of fifty-four households was ethnically mixed, comprising twenty-eight Banjarese, fourteen Javanese and twelve Malay households.

In these heterogeneous communities, the establishment of a coconut plantation is carried out individually. These villagers do not practice communal work or exchange labor with each other, as Bugis and Javanese frequently do in their native lands. The only exception to this pattern is the building of mosques or other houses of worship. Such buildings are gradually constructed through donations from community members. The mosque serves as a meeting place that unites villagers. Life in the PSF coconut region is harsh, unhealthy, unstable and risky, and religion provides migrants with psychological support. Quite often, the people are seen devotedly kneeling and bowing down to pray in assorted places. At night, elders instruct the young to read the Koran. Successful migrants who have the means may choose to make the Hajj. Some even sell their coconut plantations and boats and spend the money on the US$2,000 trip to Mecca. By 2001 five people in settlement A and seven people in settlement B had undertaken the Hajj.

Hopes for the next generation

An important reason why PSF migrants move from the coconut plantations to the villages raised on stilts is the education of their children. The migrants do not want their children to work in the PSF or grow coconuts; they hope their children can obtain the highest level of education possible. A household with such aspirations will leave the plantation to move to a village with educational facilities as soon as the children have reached school age. However, fifth or sixth grade children are a valuable source of labor. Many children from poor families have to give up education to work. In 2001, only six pupils out of sixty-four had graduated from the primary school in settlement B.

Table 12.2 shows the locations of children in settlement B who were at school in 1993. The twenty-one boys that stayed in the village assisted their parents, worked their own coconut plantations, or were employed in other jobs. Of the other twenty-six boys (who had left the village), twelve still grew coconuts, on their own or as part of a government coconut production project. Others had become construction workers in Batam, an island just opposite Singapore that has been an industrial development zone since the early 1990s, worked on a trawler, or at a lumber mill. Only three boys continued to higher education.

In 2001, sixteen unmarried girls were still at home and helped their parents with the household duties, while one helped with the family's timber-producing activities and two had moved to the houses of relatives. Only six unmarried girls had left their homes and taken up various jobs, including golf caddying and working in a supermarket in Batam. That same year, twenty-one girls had married, and fifteen remained in the settlement

Table 12.2. Occupation and whereabouts of the second generation in a peat swamp forest settlement, Sumatra

	Boys	**Girls**
Remained in the settlement		
Establishing coconut plantation	8	-
Producing boards	12	1
Higher education	1	0
Helping with housework	-	15
Housewife	-	15
Total remained in the settlement	21	31
Emigrated out		
Establishing coconut plantation	2	-
Joined government project[a]	10	-
Construction worker	3	0
Trawler worker	4	0
Sawmill worker	2	0
Worker in town	2	3
Higher education	3	0
Helping with housework	-	2
Housewife	-	11
Unknown	0	1
Total emigrated out	26	17

a: Some of the girls joined a government project following their husband, but these are not included here.

Photo 12.4. Collective logging work in 2001. Logs are sawn into boards in the forest at this time

after marriage. The husbands of these girls engage in various activities, like coconut production, working in the timber industry, fishing, or running a store. The husbands of those girls who left the village are similarly employed in miscellaneous activities, such as coconut production or working as sales clerks in the capital town. More diverse than before, the jobs of the younger generation give a glimpse of the future in this frontier community. Those that managed to go on to further education left the village and the region and did exactly what their parents expected them to do. They escaped from the frontier community and improved their career prospects elsewhere.

Discussion and conclusion

The migrant communities discussed in this chapter appear typical of a forest frontier. Both the ecological consequences of their arrival at the PSF and the reciprocal impact of the forest on the migrant's trajectory after arrival are worth highlighting. The particularly uninhabitable nature of PSF results in little concern for long-term use of the resources (because migrants plan to eventually move out and away) but also in the particular behavior of PSF migrants. The settlement structure of the pioneering communities is a result of the need to build and move along canals. This results in a pattern of frequent relocation and finally agglomeration in larger villages that provide services like health care and education. The latter is not unusual in forest frontier conditions (for example Limachi *et al.*, Chapter 3; Llanque, Chapter 4). It does, however, have its own particular manifestation in the case of PSF. For instance, the people's occupational dynamics suggest that the PSF communities remain as frontier communities until migrants abandon the PSF environment. The migrants who moved to these locations persist as migrants or strangers in the region that is now their home. The communities that eventually formed in the PSF—what we might call pseudo-communities—are a miscellaneous assemblage of migrants without a sense of solidarity. They do not develop into mature and socially stable communities, and frequently do not outlast the second or third generations.

PSF is a biologically fragile ecological system lacking resilience and stability. New migrants depend on coconuts only because coconuts allow them to make a living in the PSF environment. Commercial coconut production is directly affected by the international markets. When coconut prices fall, coconut production in a PSF environment ceases to be an option. This profoundly changes patterns of migration into and out of the PSF environment, but also how people behave after having arrived in this forest region.

The socio-cultural features of the migrants in Sumatra's PSFs distinguish them from other resident communities, a fact that constrains integration of the PSF villages. Alternatively, the different groups that inhabit this environment do have certain features in common, such as their adherence to Islam. Shared religion explains why different groups migrate to the PSF where there are already other people.

Photo 12.5. A draining canal becomes a means of transportation

The wider political context does provide its own determining factor. Indonesia's government programs promote PSF exploitation, but they do not address the needs of spontaneous migrants such as the educational and medical needs of settlements A and B discussed earlier. Rather, the government programs follow their own logic (Abe 2003). These programs are initially attractive for migrants who had already settled in the PSF, like those of settlement B. But this is mainly a result of the incentives offered (Abe 2003). It can be expected, however, that the outcomes of these government sponsored programs will not be very different from the trajectory described in this chapter. The designers of these programs fail to learn from the experiences of the people who have long settled in this environment.

Notes

1. Biologically, PSF is very vulnerable. As with other tropical forests, once deforested it is very hard to regenerate. Clearance of forest cover dries out the peat and leads to its decomposition. Dried peat is extremely flammable. During El Niño years, such as 1997, as a result of deforestation or forest degradation peat drying accelerates, and peat lands catch fire. This contributes significantly to the Southeast Asia forest fires and related smoke contamination.

References

Abe, K. 1993. Peat swamp forest in Sumatra: A perspective. *Southeast Asian Studies*, 31(3): 191–205.

Abe, K. 1997. Cari Rezeki, Numpang, Siap: The reclamation process of peat swamp forest in Riau. *Southeast Asian Studies*, 34(4): 622–632.

Abe, K. 2003. Peat swamp forest development in Indonesia and the political ecology of tropical forests in Southeast Asia. In. T.P. Lye, W. de Jong and K. Abe eds., *The Political Ecology of Tropical Forests in Southeast Asia: Historical Perspective*, 133–151. Kyoto: Kyoto University Press and Melbourne: Trans Pacific Press.

Erman, E. 1994. Tauke, Kuli dan Penguasa: Exspolitasi Hutan Panglong di Riau. In *SEJARAH: Pemikiran, Rekonstruksi, Persepsi*, 20–33. Jakarta: PT Gramesia Pustaka Utama.

Kahn, J.S. 1980. *Minangkabau Social Formation: Indonesian Peasants and the World Economy*. Melbourne, Cambridge University Press.

Pastor, G. 1929. *De Panglongs*. Weltevreden: Landsdruckkerij.

Whitmore, T.C. 1984. *Tropical Rain Forests of the Far East*. Oxford: Oxford University Press.

13
Conclusions

Wil de Jong, Lye Tuck-Po, and Abe Ken-ichi

Introduction

This volume takes a social ecology approach to the study of tropical forest migration. It proposes social ecology as a complement to human and political ecology. Where human ecology would examine migrant adaptations to environmental constraints and opportunities in the forest frontiers, political ecology would tend to take a more macro-approach in examining the creation of these frontiers. Social ecology bridges these two approaches. It explains the ecological impacts of migration and the social effects of the forest environment on migrants, as well as the macro-social consequences of migrant–frontier interaction. Social ecology is both a way of describing events, and an approach to studying them (the study of migration as a social phenomenon with multifaceted implications). Guided by this framework, this volume presents eight chapters of case studies of tropical forest migration, and three chapters that provide regional overviews for Latin America, Africa, and Southeast Asia.

Conclusions from the chapters in this book

The issues and concerns differ remarkably between the three regions reviewed here. Perz (Chapter 2) identifies the mitigating factors that help to explain migration and land-cover change in the Amazonian states of Brazil, Bolivia, Ecuador, and Peru. The migration data (rural and urban movements and the labor-force composition of migrants and non-migrants) do not adequately explain forest-cover change. Only when land-use patterns and distinct national political economies are considered can the links between migration and forest-cover change adequately be explained, for land use and political economy modify the effects of migration on land-cover change and help to explain the limited correspondence between them. From this analysis it becomes evident that factors other than migration influence land-cover change.

More detail on forest-related migration in Amazonia is provided in Limachi *et al.*'s (Chapter 3) analysis of migration into and out of the Peruvian Amazon region. Their chapter covers both a larger period and considers a wider range of political, social, and economic factors. Key factors that explain the impact of migration on forests include new economic opportunities, linked to the market booms of certain commercial forest products or agricultural crops. Migration in the Peruvian Amazon can be connected to quite a broad range of government policies, some of which were designed to influence migration, while others affected migration towards the Amazon region indirectly. The factors that explain Amazonia bound migration are pull-factors that encouraged people to move into the Amazon, and push-factors that propelled people away from their homes, especially in the case of migration from the Andes toward the Amazon since the early 1970s. The construction of roads also played a vital role, but in itself could not explain the migration and its subsequent consequences. Policies at national and regional levels have both a positive and negative effect on migration patterns. The chapter suggests that policies can be designed that encourage migrants to use land in such a way that deforestation rates decline significantly.

In Chapter 4 Llanque deals with migration in northern Bolivia, a region that remains isolated because of limited road access and depends economically on forest products such as rubber, brazil nut and timber. Llanque explicitly classifies different types of forest-related migration found in a single region: migration *into* the region; migration from rural areas to urban centers within the region; migration from urban centers to rural areas within the region; and migration between rural areas in the region. External, macro-factors influence these migration patterns. As these factors are likely to change for years to come, forest-related migration in this region will continue to change. The particular pattern of the four types of migration of this region is typical for the non-exhaustive extraction of forest products. Similar complex patterns of multidirectional migration are common for tropical regions where local resource use is linked to larger markets. Some of the major migration waves into the forest regions had little impact on the forest initially, while subsequent rural–urban migration resulted in an increased pressure on the forest. These variations are largely the result of shifts in the major economic activities of migrants, as they changed focus from forest product collection to commercial agricultural production. From this the conclusion can be drawn that migration towards a forested region in itself cannot account for increased pressure on the forests. The degree of pressure placed upon the forest ecosystem depends on the type of resource use associated with the particular migration trend, and on subsequent, more localized migration processes. Under certain circumstances movements out of the forest may imply a shift in resource use that may result in increased pressure on the forests.

Three chapters analyze tropical forest-related migration in Africa. Bradley (Chapter 5) proposes a number of generic situations and processes. An important contribution of this review is its focus on history and the effect of past events on current dynamics. Bradley questions the quality of data used to explain processes of change in history. This is particularly true for data on the extent of forest cover and the rate and nature of deforestation. Recent studies have cast considerable doubt on what were once accepted as truths. For the future, there are positive signs that co-management of forest resources, in which local people become stakeholders of valuable resources, may go some way to securing a long-term future for forest resources, although this is only ever likely to affect small areas. Any gains in forest cover through co-management, however, are likely to be threatened by global patterns of resource consumption. Mining and timber interests, the expanding bush-meat trade with its global, if informal, reach, and the continuing impoverishment of the peasantry forced to seek out new and better land, do not bode well for a sustainable future.

In Chapter 6 Acheampong contributes evidence for this and other arguments in the volume based on an analysis of migration into the Western Region of Ghana. This is one of the remaining forest frontiers of the country. Commercial cocoa production has been a key element in this migration, and the impact that immigrants have had on various aspects of rural life in the forest frontier zone has been significant. However, cocoa production is only one factor that explains the impact of migration into this region. The cultural makeup of locals and migrants, the local biophysical conditions, government economic policies, and the world market for cocoa and other products are all contributory factors. The arrival of migrants has changed social relations of production and land tenure systems. It has facilitated the alienation of land and labor and the process of social differentiation, resulting in the emergence of large landlords, absentee landowners with multiple holdings, and landless peasants and smallholder cultivators with insufficient land to farm. Key contributing changes linked to migration are changes of ethnic composition at the forest frontier, the distribution of land ownership, the evolving tenure systems, and variations in the resource use patterns of current residents.

Dounias and Leclerc (Chapter 7) draw a link between the African and Asian continents, discussing how the contemporary behavioral patterns of two remote hunter-gatherer societies, Punan of Borneo and Pygmies of

Central Africa, need to be explained by modern and pre-modern migration patterns. The authors turn to the concept of historical ecology applied at a landscape scale to assess the complex and causal interactions between nature and culture and to suggest directions of change. This approach can help to understand and address how environmental degradation threatens these forest dwelling groups. Changes in subsistence strategies, demography, or perception have through time resulted in both the intentional and unintentional modification of the forests. Given the rapid deterioration of tropical forests, it is extremely important that we draw on our knowledge of historical events that influence the migration of hunter-gatherer societies, and subsequently affect their contemporary exploitative and managerial practices.

Potter, in Chapter 8, reviews migration in Southeast Asian forests and also emphasizes the historical dimension of the phenomenon. Migration to frontiers in the Philippines, Indonesia, Vietnam and Thailand has boomed since the early 1950s. Migration patterns have had many different causes, and these causes have changed over time. They coincided with new economic opportunities, but were also part of government policies. The need to move to forest zones out of desperation for new livelihood opportunities seems to be a fairly recent phenomenon. Different types of migration, however, have had and are having different levels of impact on the forest. Where migration leads to forest product extraction, the impact may be minimal. Even when people engage in agriculture at the forest frontier, their impact is a function of what exactly they do, and how long they stay. The chapter synthesizes three explanatory frameworks: the neo-classical, neo-Malthusian, and political ecology frameworks and adds a new one, social ecology. This approach allows the inclusion of factors like unfamiliarity with the frontier environment, the cultural and social traits of the in-migrants, and cultural differences between migrants and local groups, in explaining how immigrants fared in their new area of settlement, and how that translated into forest conversion.

Andaya in Chapter 9 presents the various factors that explain early modern migration patterns in Asia. They include cultural factors, for instance in the case of the Iban of Borneo, who migrated to pursue rank enhancement through founding new settlements and conquering the forest. The greatest penetration into and intrusion on the forest has been the result of economic development. In the early modern period, world demand for pepper and tin encouraged states and individuals to clear the forest and to use its resources to facilitate the planting of pepper and the extraction of tin. Large forested areas were beginning to be cleared in parts of Sumatra, the Malay Peninsula, and Borneo for the planting of pepper. Equally intrusive were the movements of laborers, both free and enslaved, to work these pepper gardens. The demand on the soil's nutrients placed by the pepper plant resulted in the steady clearing of new land to maintain productivity. Tin did not require as much cleared space, but the introduction of Chinese methods of tin extraction and the establishment of Chinese mining settlements had a major impact on the forest. As the forests of Southeast Asia become depleted, there is a corresponding diminishment of their mystery and hence their awe. They are now regarded as the last frontier to be economically exploited by the few, rather than a resource which needs to be preserved for the wellbeing of the whole community. Those who persist in maintaining a lifestyle dependent on the forests will be "encouraged" to leave their forest domain and become a progressive and modern member of the national society.

In Chapter 10 Tran Van Con addresses forest-based migration into Vietnam's Central Highlands since the mid-1990s. He argues that the causes and effects of migration to places like the Central Highlands must be seen from different angles, in different temporal perspectives, and at different levels of geographical resolution. He uses this analysis as a contribution to planning strategies to tackle this issue, and formulates a number of relevant recommendations. Key recommendations are the need for multi-stakeholder collaboration, better incorporation of customary forest management and tenure, and a better consideration of the subsistence needs of rural dwellers for forestland. He recommends that forest policies and official programs should strengthen community participation in forest management and increase direct benefits from forest production activities.

Eghenter (Chapter 11) analyzes a special case of emigrants who return to the Apo Kayan, a forested region in East Kalimantan, Indonesia, to collect eaglewood. The critical issues are membership in the group and related rights to natural resources in the territory. Membership becomes unclear in the case of return migrants or outsiders who had married into the community. People by virtue of belonging to the same ethnic and family group may be "insiders", and as such may legally claim a share of the resources and land in the village of origin. However, according to local rules, they relinquished claims when they moved out. No unanimous policy has been proposed on this issue. This conundrum has led to a weakening of the natural resource management practices that prevented overexploitation in the past. The economic situation of the Apo Kayan was also negatively affected as the eaglewood boom created a "one-product commodity" economy with unstable market prices for all goods. This instability led to a social shift in the communities, and those who used to hold power based on greater wealth and related village political influence often lost control. Divisions in the communities, the arrival of outsiders, and strategies of "imposed" social and economic obligations practiced by the traders negatively affected the potential for cooperation in the community. A one-commodity economy and unequal economic opportunities continue to undermine the long-term security and wellbeing of the local people. Fuzzy "membership" generates more ambiguity and contradictions in the local definitions of ethics of access.

In Chapter 12 Abe narrates the fate of Bugis who have moved to peat swamp forests in Sumatra, Indonesia from the late 1940s onwards. This chapter, like chapters before, demonstrates the determining outcome of the environment, the cultural attributes of the migrants and residents in the region of migration, and the wider political-economic constellation of the Indonesian State. The result is a quite unusual pattern of both occupation and subsequent relocation. The new settlers into the Sumatran peat swamp forest carve out a harsh existence in a hostile environment, where little other than coconut palms can be grown. The environmental features impose a spatial occupation that limits social interaction and access to services. Many new residents into this region, therefore, move to regional towns after establishing their coconut estates. They become non-resident farmers as they return to their coconut estate only to harvest and engage in minimal tending. A large number of the children of the Sumatran peat swamp immigrants move on to other occupations.

Migration and forest-cover change to come

Potter (Chapter 8) suggests that the forest frontier era and poverty-related forest frontier immigration is reaching its final stage. We need to reflect on this statement when considering the future of tropical forest-related migration. A speculative answer would be that this is indeed, and increasingly so, the case in many Asian countries, and to some extent also in Latin America. It is yet, however, quite unpredictable what will happen in the West African tropical forest region. Even if this region stabilizes politically, it will still not have economies that adequately support livelihoods and governments that can adequately control land use. Hence, the forest frontier phenomenon, driven by people who are looking for agricultural land, pushed away by lack of opportunities, climatic catastrophes, or violence elsewhere, may still be a factor in significant parts of Africa.

Even though poverty and land-driven forest frontier migration in Latin America may be of the past, forest-related migration will be especially influenced by plans to integrate the region through a vastly expanded network of paved roads. One road network that is likely to have a large influence on the regions discussed by Perz (Chapter 2), Llanque (Chapter 4), and Limachi *et al.* (Chapter 3) is the Inter-Oceanic Highway that is to connect the Pacific and Atlantic oceans, running from Peru, through its southern tropical rainforest area, passing very close by the northern Bolivian forest region that Llanque discusses, going through Acre in Brazil and onward to the Atlantic coast (see http://map.amazonia.net). In 2004, President Lula from Brazil visited his colleague Alejandro Toledo in Lima, Peru. They signed sixteen agreements, one of which was to finish building this road within the next two years. This, and other roads of the South American network, including roads inside the vast Brazilian Amazon, will trigger important new waves of migration.

Other smaller migration booms will continue to happen, even in places where the forest frontier has disappeared. The road aside, northern Bolivia is already attracting large numbers of people because of the increasing cross-border trade and government policies (Llanque, Chapter 4; Perz, Chapter 2). The region where the Kalimantan provinces of Indonesia meet with Malaysia's Sarawak and Sabah states experiences similar processes. The region shared by China, Vietnam, and Laos is increasingly integrating by a similar road system. Again migration in such regions will be triggered by economic opportunities, new roads, or national and international economic fluctuations.

The nature of social ecological inquiry will change as a result of the above factors. A more systematic assessment of interactions at the destinations of forest-bound migrants will continue to be a focus of inquiry, although the type of actors and their social contexts may change. There may be less need to understand the constraints of the migrants who come with basic, unfulfilled needs for land and livelihood than the strategies of those who come to benefit from opportunities, and how the latter are linked to national and possibly local economic, environmental, and social goals. Interactions between entrepreneurial migrants and local residents, each with their own cultural and social identities, will likely become an increasingly important topic.

This picture, however, might change totally if factors such as climate change or international security start playing a more significant role. The implications for tropical forest-related migration, and for social ecology's modes of inquiry, are yet hard to predict.

Index